THE NEW FRONTIERS OF GENETICS

GENE EDITING, CRISPR AND THE FUTURE OF HUMANITY

DAVID SANDUA

The new frontiers of genetics: Gene editing, CRISPR and the future ɾ
humanity.

Kindle Direct Publishing
Paperback Edition 2023

THE NEW FRONTIERS OF GENETICS

GENE EDITING, CRISPR AND THE FUTURE OF HUMANITY

DAVID SANDUA

"As we rewrite genetic code, we're authoring the prologue to a future where genetic diseases are but a memory."

Dr. Sophia Lewis, Gene Therapist.

INDEX

8

I. INTRODUCTION

The world of genetics has always been a captivating and tantalizing subject, continuously pushing the boundaries of human understanding and paving the way for unprecedented advancements. In recent years, one breakthrough in particular has revolutionized the field: gene editing, made possible by the groundbreaking tool known as CRISPR. This powerful technology has ushered in a new era of possibilities, enabling scientists to manipulate the very fabric of our genetic code with unparalleled precision. The potential of gene editing and CRISPR to transform medicine, agriculture, and even alter the future of humanity is nothing short of astounding. In this essay, we will embark on a journey to explore the exciting frontiers of genetics, delve into the intricacies of gene editing, and contemplate the profound implications it holds for our society. As our understanding of genetics has deepened over time, the notion of altering DNA to cure diseases and enhance human capabilities has captured the imaginations of scientists and the general public alike.

CRISPR has emerged as a game-changer in this arena. This revolutionary genetic engineering tool acts as a pair of molecular scissors, capable of precisely cutting and modifying DNA. Unlike previous methods of gene editing, CRISPR offers a level of accessibility and efficiency that was once unfathomable. It allows researchers to easily target and edit specific genes, opening up a world of possibilities for treating hereditary diseases, such as cystic fibrosis or sickle cell anemia, that were once deemed incurable. The potential of gene editing extends far beyond the boundaries of medicine. With CRISPR, scientists have the power to reshape entire ecosystems by modifying the genetic makeup of organisms. This technology has the potential to design crops that are resistant to pests and diseases, increasing yields and reducing the reliance on potentially harmful pesticides. It could also enable the creation of genetically modified organisms that can tolerate extreme environmental conditions, such as drought or high temperatures, potentially revolutionizing agriculture and ensuring food security in a changing climate. CRISPR has opened up avenues for the development of new biofuels, renewable materials, and even more efficient methods of waste disposal. While the possibilities of gene editing are undeniably exciting, they are not without ethical and societal considerations. As we delve deeper into the realm of manipulating our genetic blueprint, questions of safety, equity, and unintended consequences arise. Ensuring the ethical use of gene editing technologies is crucial to prevent potential abuses and maintain public trust. Issues of equity and access must be addressed, as the benefits of gene editing should not be limited to a privileged few. The long-term effects of altering genes in organisms and potentially spreading those changes through reproduction or environmental interactions require careful consideration and thorough risk assessments. Gene editing and CRISPR have undoubtedly brought us to the edge of science, but what lies beyond? The implications of these breakthroughs for the future of humanity are profound and multifaceted. On one hand, the ability to eliminate genetic diseases and improve overall health and longevity holds tremendous value for individuals and society as a whole. It offers hope for a future where debilitating conditions can be eradicated, and human potential can be maximized. On the other hand, the power to manipulate genes raises complex ethical questions about what it means to be human, the boundaries of natural selection, and the potential for eugenics. The advent of gene editing and CRISPR has revolutionized the field of genetics and opened up new frontiers of possibility. With the power to precisely modify DNA, these breakthroughs hold immense potential for curing genetic diseases, reshaping ecosystems, and possibly even altering the very essence of what it means to be human. Along with these exciting prospects come ethical and societal considerations that must not be overlooked. As we embark on this journey to the edge of science, we must navigate carefully and thoughtfully, ensuring that the future of gene editing aligns with our values and aspirations as a global society.

BACKGROUND ON GENETICS

Genetics, the study of heredity and the variation of inherited characteristics, is a field that has fascinated scientists and researchers for centuries. Dating back to Gregor Mendel's experiments with pea plants in the 19th century, genetics has sought to unravel the intricacies of life itself. It was through Mendel's work that the fundamental principles of genetics, such as dominant and recessive traits, were established. From these humble beginnings, genetics has evolved into a complex and multidisciplinary field that encompasses various sub-disciplines such as molecular genetics, population genetics, and functional genomics. With advancements in technology and the advent of gene editing tools like CRISPR, the field of genetics has witnessed a revolutionary breakthrough that has the potential to reshape the future of humanity. Deoxyribonucleic Acid (DNA), commonly referred to as the blueprint of life, is at the core of genetics. DNA carries the genetic information that determines the traits of an organism, from its physical attributes to its susceptibility to certain diseases. The discovery of the structure of DNA by James Watson and Francis Crick in 1953 opened the door to the field of molecular genetics. Scientists soon realized that by understanding the structure and function of DNA, they could unravel the secrets of life at the most basic level. One of the most significant milestones in the field of genetics was the completion of the Human Genome Project (HGP) in 2003. This massive international effort aimed to determine the sequence of nucleotide base pairs that make up human DNA and identify all the genes present in the human genome. The HGP not only provided researchers with a comprehensive map of the human genome but also paved the way for further research in the fields of personalized medicine and gene therapy. By identifying specific genes associated with diseases, researchers could develop targeted therapies and treatments for individuals based on their genetic makeup.

The advent of gene editing technologies, particularly CRISPR (Clustered Regularly Interspaced Short Palindromic Repeats), has revolutionized the field of genetics in recent years. CRISPR-Cas9 is a revolutionary gene editing tool that enables scientists to precisely edit DNA sequences in living organisms. It functions by utilizing a guide RNA molecule to target a specific DNA sequence, and an enzyme called Cas9 to cut and modify the DNA at that site. CRISPR has the potential to revolutionize medicine, agriculture, and even environmental conservation.

In the realm of medicine, CRISPR holds immense promise for the treatment of genetic disorders. By targeting disease-causing genes, researchers can potentially correct the underlying genetic defects, offering a cure where none existed before. For example, researchers have successfully used CRISPR to correct the mutation responsible for a form of hereditary blindness in mice, providing hope for similar treatment in humans. CRISPR has shown potential in the field of cancer research, with studies indicating its ability to selectively kill cancer cells while leaving healthy cells unharmed. Beyond disease treatment, CRISPR also has the potential to reshape the future of agriculture. By modifying the genes of crops, researchers can enhance their nutritional value, increase their resistance to pests and diseases, and even improve their tolerance to environmental conditions such as drought. This could have significant implications for global food security, as genetically modified crops could potentially yield higher agricultural productivity and reduce the need for chemical pesticides and fertilizers. CRISPR could play a vital role in environmental conservation by assisting in species conservation efforts. Scientists have already begun exploring the use of gene editing to address issues such as the decline of endangered species and the eradication of invasive species. The ability to manipulate the genetic composition of organisms could provide novel approaches to conserving biodiversity and restoring ecosystems that have been disrupted by human activities.

Genetics and the groundbreaking gene editing technology of CRISPR have the potential to revolutionize various aspects of human life, from medicine to agriculture and environmental conservation. The field of genetics, with its ability to understand and manipulate the building blocks of life, holds a remarkable future for humanity. As researchers continue to push the boundaries of what is possible, it is important to navigate ethical considerations and ensure responsible use of these powerful tools to harness the full potential of genetics for the greater good of humanity.

INTRODUCTION TO CRISPR TECHNOLOGY

CRISPR (Clustered Regularly Interspaced Short Palindromic Repeats) technology has emerged as a revolutionary tool in the field of genetics, holding the potential to transform various aspects of our lives. With its ability to precisely edit DNA sequences, CRISPR has brought about a new frontier in gene editing, opening up endless possibilities for advancements in medicine, agriculture, and even the preservation of endangered species. This technology, derived from a naturally occurring immune system found in bacteria, has captured the attention of scientists and researchers worldwide for its simplicity, efficiency, and versatility. The rapid development and widespread adoption of CRISPR technology have allowed researchers to explore and manipulate the building blocks of life in unprecedented ways, heralding a new era of genetic engineering that can shape the future of humanity. From curing genetic diseases to modifying organisms, CRISPR has proven to be a powerful tool that holds immense promise in addressing some of the most pressing challenges facing humanity today. One of the most significant applications of CRISPR technology is in the field of medicine. Through the targeted editing of DNA, scientists can potentially correct genetic mutations that cause various diseases, offering the hope of curing previously untreatable conditions. By using CRISPR to edit the genetic code of diseased cells, researchers can theoretically eliminate the roots of diseases, rather than merely treating their symptoms. This breakthrough has the potential to revolutionize the treatment of genetic disorders such as sickle cell anemia, cystic fibrosis, and

Huntington's disease, which have long plagued individuals and their families. The ability to edit the genetic code also opens doors to personalized medicine, whereby treatments can be tailored to an individual's unique genetic makeup, leading to better therapeutic outcomes and reduced side effects.

In addition to its medical applications, CRISPR technology has the potential to transform agriculture and address the challenges of food security. With the world's population projected to reach 9.7 billion by 2050, scientists are under increasing pressure to find sustainable ways to feed a growing global population.

CRISPR offers a promising solution by enabling the development of genetically modified crops that are more resilient to pests, diseases, and environmental stressors. By editing specific genes responsible for traits such as drought resistance or nutritional content, scientists can create crops that can thrive in adverse conditions and provide higher yields. This technology also holds the potential to improve the nutritional value of crops, enhancing access to essential vitamins and minerals in regions where malnutrition is prevalent. It is imperative to ensure that the use of CRISPR in agriculture is regulated carefully to address potential ethical concerns and to prevent unintended consequences.

CRISPR technology has the potential to play a crucial role in conservation efforts and the preservation of endangered species. As human activities continue to disrupt ecosystems and drive species to the brink of extinction, conservationists are grappling with the challenge of protecting biodiversity.

Through the application of CRISPR, scientists can potentially revive populations of endangered species by altering their genetic makeup to increase their chances of survival. This could involve editing the genes responsible for susceptibility to diseases, enhancing reproductive capabilities, or even reintroducing traits that have been lost due to human interference. Nonetheless, the use of CRISPR technology in conservation raises complex ethical questions and requires careful consideration to ensure that it is employed responsibly and with proper regard for the long-term ecological implications. CRISPR technology represents a significant breakthrough in genetics, propelling us into a world with unimaginable possibilities. With its ability to precisely edit DNA, this revolutionary tool holds the potential to cure genetic diseases, address food security challenges, and even reshape ecosystems. Nonetheless, the practical implementation of CRISPR technology requires careful consideration of ethical, legal, and social implications. As scientists continue to push the boundaries of gene editing, it is crucial to maintain a balanced approach that upholds ethical principles, fosters transparency, and promotes responsible use. The future of CRISPR technology holds immense promise, but its success will ultimately depend on how we navigate the complex terrain of genetic engineering, ensuring that our actions advance the well-being of humanity and the natural world.

THESIS STATEMENT

The advent of gene editing, particularly CRISPR, holds immense potential in curing diseases, altering organisms, and consequently reshaping the future of humanity.

The advent of gene editing, particularly CRISPR, holds immense potential in curing diseases, altering organisms, and consequently reshaping the future of humanity. In terms of disease eradication, CRISPR has demonstrated its ability to disrupt disease-causing genes and potentially provide a cure for genetic disorders. By precisely targeting and modifying specific genes associated with these disorders, CRISPR has the potential to revolutionize the field of medicine. In fact, researchers have already made significant progress in using CRISPR to treat diseases such as sickle cell anemia and cancer. For instance, in a study published in the journal Science, researchers successfully used CRISPR to edit a gene in human embryos that causes hypertrophic cardiomyopathy, a heart disease that affects millions of people worldwide. The potential of CRISPR in curing diseases extends beyond genetic disorders to infectious diseases as well. By editing genes in disease-causing organisms, CRISPR could potentially prevent the spread of infectious diseases such as malaria or Zika virus, saving countless lives in the process.

In addition to its potential in disease eradication, gene editing using CRISPR has the power to alter organisms in ways that were previously unimaginable. With precise gene editing tools like CRISPR, researchers have the ability to modify the DNA of plants, animals, and even humans, opening up new possibilities for

21

improving agriculture, creating new breeds of crops, and enhancing the human species. In agriculture, CRISPR can be used to make crops more resistant to diseases, pests, and environmental challenges, potentially revolutionizing food production and addressing the growing global food crisis. Similarly, in the field of animal breeding, CRISPR can be used to create genetically modified organisms with enhanced traits, such as increased milk production in cows or disease resistance in pigs, leading to improved livestock farming practices and increased efficiency. The ability to modify the human genome using CRISPR raises ethical questions and discussions about the potential of creating "designer babies" with desirable traits. While the idea of genetically engineering humans may be controversial, CRISPR has the potential to eliminate genetic diseases and ensure a healthier future for generations to come. The implications of gene editing and CRISPR technology on the future of humanity are profound. As we harness the power of gene editing, we are entering a new era where we have the ability to shape our own biology. This has significant implications for the future of healthcare and human well-being. The ability to edit genomes could lead to personalized medicine tailored to an individual's specific genetic makeup, thereby increasing the efficiency and effectiveness of treatment. The potential to extend the human lifespan and enhance cognitive abilities through genetic modifications raises important ethical considerations about the implications of "playing god" and the potential for widening social inequalities.

While the possibilities presented by gene editing and CRISPR are undoubtedly exciting, it is crucial to tread carefully and address the ethical, legal, and social implications that arise from these technologies. The ability to edit genes raises pressing questions about the moral responsibility of scientists and the limits of our intervention in the natural order. Issues such as accessibility, affordability, and equity in the access to gene-editing technologies must be considered to ensure that these breakthroughs benefit all of humanity and do not exacerbate existing inequalities.

The advent of gene editing, particularly CRISPR, holds immense potential in curing diseases, altering organisms, and reshaping the future of humanity. The ability to edit and modify genes has already shown promising results in disease eradication and holds the potential to revolutionize the fields of medicine, agriculture, and human genetics. As we navigate these new frontiers of genetics, we must carefully navigate the ethical, legal, and social challenges that arise, to ensure that these technologies are harnessed for the benefit and betterment of all of humanity.

Gene editing and CRISPR technology have opened up a whole new world of possibilities in the field of genetics. This exciting realm of scientific discovery has the potential to cure diseases, change organisms, and reshape the future of humanity. The ability to edit genes and manipulate DNA sequences gives scientists the power to make precise changes in the genetic code, which was once thought to be impossible. This newfound ability has the potential to revolutionize medicine, agriculture, and even our own evolution as a species. One of the most promising applications of gene editing is in the realm of disease treatment and prevention. With CRISPR technology, scientists can target and disable disease-causing genes, effectively curing genetic disorders that were once thought to be incurable. This breakthrough has the potential to save countless lives and improve the quality of life for those who suffer from genetic diseases. Diseases such as cystic fibrosis, sickle cell anemia, and Huntington's disease may one day be completely eradicated through gene editing techniques. In addition to curing genetic diseases, gene editing can also be used to prevent the transmission of genetic disorders from one generation to the next by editing the reproductive cells of individuals with these disorders. This breakthrough has the potential to eliminate certain genetic disorders from populations altogether. Beyond disease treatment, gene editing also holds great promise for improving agricultural practices and feeding a growing global population. By manipulating the genes of crops and livestock, scientists can create organisms that are more resistant to pests, diseases, and harsh environmental conditions. This has the potential to increase crop yields and improve the efficiency of agricultural production. Gene-edited crops could also be engineered to have enhanced nutritional profiles, providing essential vitamins and minerals to populations that are lacking in these nutrients. In a world plagued by food insecurity and climate change, gene editing technologies have the potential to revolutionize our ability to feed the planet in a sustainable and efficient way. Perhaps the most controversial and ethically complex application of gene editing is in the realm of human enhancement and modification. With the ability to edit human genes, the possibility of altering physical and cognitive abilities becomes a reality. While this may sound like the stuff of science fiction, the reality is that gene editing has the potential to reshape the future of humanity in unprecedented ways. It raises questions about what it means to be human, the limits of nature, and the ethics of manipulating our own genetic makeup.

On one hand, gene editing could be used to eliminate genetic predispositions to certain diseases and disorders, improving the overall health and well-being of individuals. This could lead to longer and healthier lives for future generations. On the other hand, the ability to enhance physical and cognitive abilities raises concerns about creating a genetic elite and exacerbating existing social inequalities. It also raises ethical questions about the potential for designer babies and the commodification of human life. These are complex ethical dilemmas that will need to be carefully considered and addressed as gene editing technologies continue to advance. Gene editing and CRISPR technology hold tremendous potential for improving human health, feeding a growing global population, and reshaping the future of humanity. The ability to edit genes and manipulate DNA sequences has the power to cure diseases, improve agricultural practices, and even enhance and modify human abilities. With this power comes great responsibility. Ethical considerations must be at the forefront of these advancements, as the potential to create a genetic elite and the commodification of human life raises important ethical questions. As we journey to the edge of science, we must proceed with caution and careful consideration of the implications and impacts of these new frontiers of genetics.

II. UNDERSTANDING GENETICS

In order to fully grasp the significance of gene editing and CRISPR technology, it is essential to have a thorough understanding of genetics. Genetics is the branch of biology that studies genes, traits, and heredity in living organisms. Genes, which are made up of DNA, are the fundamental units of heredity and determine the characteristics and traits of an organism. They are responsible for the inherited traits we see, such as eye color, hair color, and height. Genes also play a crucial role in the development of diseases. Certain mutations or variations in genes can result in genetic disorders or an increased susceptibility to certain conditions. Understanding the structure and function of genes is essential for geneticists to diagnose and treat genetic diseases.

At the heart of the revolutionary CRISPR technology is the ability to precisely edit genes. The CRISPR system is a powerful gene editing tool that allows scientists to make targeted changes to the DNA of living organisms. CRISPR-Cas9, the most well-known and commonly used technique within the CRISPR system, involves using an enzyme called Cas9 to cut the DNA at a specific location, and then introducing a small piece of DNA called a guide RNA that acts as a template for repairs. This process allows scientists to add, remove, or modify specific sections of DNA, thereby altering the function of genes. One of the major implications of gene editing and CRISPR technology is the potential to cure genetic diseases. By identifying and correcting the specific genetic mutations responsible for a disease, scientists can potentially eliminate the root cause of the disorder. This has already been demonstrated in a number of studies, with researchers successfully using CRISPR to correct mutations in genes associated with various genetic conditions, such as sickle cell anemia and cystic fibrosis. In the future, as our understanding of genetics and gene editing techniques continues to improve, it is possible that we may be able to effectively treat a wide range of genetic disorders that were previously thought to be incurable.

CRISPR technology opens up new possibilities in the field of agriculture and food production. Through gene editing, scientists can develop crops that are more resistant to pests, diseases, and environmental factors, thereby increasing crop yields and food security. Gene editing can be used to enhance the nutritional content of crops, leading to more nutrient-rich foods. This has the potential to address the issue of malnutrition and improve the overall health of populations in developing countries. It is important to carefully consider the ethical implications of gene editing in agriculture, as the potential unintended consequences and long-term effects on ecosystems are still not fully understood. In addition to curing diseases and improving food production, gene editing and CRISPR technology also have the potential to reshape the future of humanity in ways we have never imagined. With the ability to manipulate the genetic makeup of organisms, it is now possible to create genetically modified organisms (GMOs) with desirable traits. This opens up possibilities for creating designer babies, where parents can select specific traits and characteristics for their children. While this idea may sound like science fiction, it is a real possibility that raises ethical concerns about the potential for creating a genetically divided society and the potential for unintended consequences.

Genetics is a fundamental field of study that underlies our

understanding of traits, diseases, and heredity. The advent of CRISPR technology has given us the ability to precisely edit genes and hold great promise for curing genetic diseases, enhancing agriculture, and reshaping the future of humanity. With such power comes great responsibility, and it is essential that we engage in thoughtful and ethical discussions about the implications and potential risks of gene editing. As we continue to explore the new frontiers of genetics, we must strive to use these breakthroughs for the betterment of humanity while being mindful of the long-term consequences.

BRIEF HISTORY OF GENETICS

The field of genetics dates back thousands of years, with early roots in the domestication of plants and animals. It was not until the mid-19th century that Gregor Mendel laid the foundation for modern genetics with his experiments on pea plants. Mendel discovered the basic principles of inheritance, including the concept of dominant and recessive traits, as well as the idea of genetic variation. His work went largely unnoticed at the time, but it would later become the cornerstone of modern genetics.

The next major breakthrough in the field of genetics came in 1953 with the discovery of the structure of DNA. James Watson and Francis Crick famously described DNA as a double helix, a twisted ladder-like structure composed of nucleotides. This discovery was instrumental in understanding how genetic information is stored and passed on from generation to generation. It also paved the way for further advancements in genetics research. In the decades that followed, scientists began to unravel the complex mechanisms of genetic inheritance. They discovered the role of genes in determining physical traits, and how mutations in these genes can lead to genetic disorders. The development of techniques like polymerase chain reaction (PCR) allowed scientists to amplify and study specific segments of DNA, further enhancing our understanding of genetic variation and disease.

Fast forward to the present day, and we find ourselves on the cusp of a genetic revolution. The development of CRISPR technology has revolutionized the field of genetics. CRISPR allows scientists to edit genes with unprecedented precision, opening up a wealth of possibilities in biomedical research and beyond.

CRISPR works by harnessing the power of a bacterial immune system. Bacteria use CRISPR to recognize and destroy viral DNA, effectively protecting themselves from infection. Scientists have adapted this system to target and edit specific genes in a variety of organisms, including humans. By using a molecule called RNA, which acts as a guide, scientists can direct an enzyme called Cas9 to a specific location in the DNA. Cas9 then cuts the DNA at that location, allowing researchers to delete, add, or modify genetic material. The potential applications of CRISPR are vast and varied. In the field of medicine, CRISPR holds promise for the treatment of genetic disorders. By correcting mutations in the DNA, scientists hope to cure diseases that were once thought to be untreatable, such as cystic fibrosis and sickle cell anemia. CRISPR has already shown success in treating genetic disorders in animal models, and clinical trials are underway to test its effectiveness in humans. Beyond medicine, CRISPR has the potential to transform agriculture by creating crops that are more resistant to pests and diseases, and that have increased nutritional value. It could also be used to engineer biomaterials, create renewable energy sources, and even revive extinct species. The possibilities are truly staggering. The ethical implications of gene editing are also profound. The ability to manipulate the genetic code raises questions about the boundaries of human intervention in the natural world. Should we be playing the role of gods, altering the fundamental building blocks of life? And who gets to decide which genes are edited and for what purposes? These are questions that society must grapple with as technology continues to advance. The field of genetics has come a long way since its humble beginnings. From Gregor Mendel's experiments with pea plants to the discovery of the structure of DNA, our understanding of genetics has evolved rapidly. The advent of CRISPR technology has opened up new frontiers in genetic research, offering unprecedented opportunities to cure diseases, modify organisms, and reshape the future of humanity. With great power comes great responsibility, and it is up to us to navigate the ethical challenges that accompany these new advancements. The path ahead is both exciting and uncertain, but one thing is clear: the world of genetics is on the brink of a revolution.

GENETIC TRAITS AND INHERITANCE

Understanding the complexities of genetic traits and inheritance is crucial to fully comprehend the potential impact of gene editing and CRISPR technologies on the future of humanity. Genetic traits are fundamental characteristics that are determined by the genes we inherit from our parents. These traits can range from easily observable physical attributes, such as eye color or height, to more abstract traits like intelligence or predisposition to certain diseases. The hereditary nature of these traits is what makes them fascinating and worth exploring. Inheritance, on the other hand, refers to the passing on of genetic information from one generation to the next. The basic unit of inheritance is the gene, which is a segment of DNA that contains instructions for building proteins. These proteins play a vital role in the functioning and development of our bodies. In the process of inheritance, genes are transferred from parents to offspring through sexual reproduction. In humans, the inheritance of traits follows certain patterns. The most commonly studied inheritance pattern is known as Mendelian inheritance, named after the pioneering geneticist Gregor Mendel. Mendelian traits are those that follow predictable patterns of inheritance, such as dominant or recessive traits. For example, if both parents have brown eyes, it is highly likely that their offspring will also have brown eyes. If one parent has blue eyes and the other has brown eyes, the offspring may have either brown or blue eyes, depending on the dominant and recessive genes involved. In addition to Mendelian traits, there are other types of inheritance patterns that involve multiple genes and environmental factors. These complex traits, such as height or intelligence, are influenced by the interplay of multiple genes as well as environmental factors like nutrition, lifestyle, and exposure to various stimuli. The study of these complex traits and their inheritance patterns is a fascinating and ongoing area of research. The discovery of CRISPR technology has revolutionized our ability to manipulate and edit genes, opening up whole new possibilities for understanding and altering genetic traits. CRISPR, refers to a naturally occurring system in bacteria that acts as an immune defense mechanism against viral infections. Scientists have harnessed this system to develop a powerful gene editing tool that allows for precise and efficient modifications to the DNA of living organisms. CRISPR works by utilizing a molecule called RNA, which is capable of guiding a protein to a specific section of DNA. This protein, known as Cas9, acts as a pair of molecular scissors, cutting the DNA strand at the desired location. The cell's natural repair mechanisms then take over to either repair the cut DNA or introduce desired changes by inserting a new piece of DNA. With the help of CRISPR technology, scientists can now edit specific genes, remove harmful mutations, or introduce new traits into an organism's DNA.

The potential applications of CRISPR are vast and can be both promising and controversial. On one hand, gene editing holds the promise of curing genetic diseases by correcting faulty genes. This could mean eradicating devastating genetic disorders such as cystic fibrosis, sickle cell anemia, or Huntington's disease. In addition to disease prevention, CRISPR technology also offers the potential for enhancing desirable traits, such as increasing crop yields or creating disease-resistant livestock. The ethical implications of gene editing are still being debated, particularly when it comes to modifying the human germline, which would affect future generations. A comprehensive understanding of genetic traits and inheritance is crucial to fully comprehend the potential of gene editing and CRISPR technologies. The inheritance of genetic traits follows predictable patterns, albeit with complexities and variations involved in complex traits. The discovery of CRISPR has opened up new frontiers in gene editing, allowing for precise modifications to DNA. While the potential applications of CRISPR are promising, ethical considerations and ongoing research are necessary to harness its potential responsibly and ethically. The future of humanity is thus at the intersection of genetics, CRISPR, and mindful decision-making.

ROLE OF DNA IN GENETICS

In the field of genetics, DNA plays a central role in determining the traits and characteristics of living organisms. Deoxyribonucleic acid, commonly known as DNA, is a complex molecule that carries the genetic information necessary for the development and functioning of all known living organisms. It is made up of a sequence of nucleotides that contain the instructions for building and maintaining an organism. DNA is found within the nucleus of a cell and is organized into structures known as chromosomes. Each chromosome contains many genes, which are specific segments of DNA that code for particular proteins. These proteins are responsible for carrying out the various functions of the body and ultimately determine an organism's characteristics.

One of the most significant contributions of DNA to genetics is its ability to undergo replication and transcription, allowing for the transmission of genetic information from one generation to the next. When a cell divides, its DNA must be replicated so that each daughter cell receives an identical copy of the genetic material. This process ensures that the genetic information is passed on and inherited by subsequent generations. Replication is achieved through the complementary base pairing of the nucleotides in DNA. Adenine pairs with thymine, and cytosine pairs with guanine, forming the famous double helix structure of DNA.

Another crucial role of DNA in genetics is the process of transcription, which involves the synthesis of a complementary RNA molecule from a DNA template. This process occurs in the nucleus and is catalyzed by an enzyme called RNA polymerase. During transcription, the DNA double helix unwinds, exposing a specific gene sequence. RNA polymerase then reads the DNA sequence and synthesizes a complementary RNA molecule by adding nucleotides that are complementary to the exposed DNA template strand. This RNA molecule, known as messenger RNA (mRNA), carries the genetic information from the DNA to the ribosomes in the cytoplasm, where it acts as a template for protein synthesis. The importance of DNA in genetics is further highlighted by its role as the carrier of genetic information, or genes, that determines an organism's traits and characteristics. Genes are specific sequences of DNA that code for particular proteins. These proteins are the building blocks of cells and perform a wide range of functions in the body. They are responsible for traits such as eye color, height, and susceptibility to certain diseases. The expression of genes is regulated by a complex network of molecular interactions that control when and where certain genes are turned on or off. This regulation allows for the development and specialization of different cell types in an organism, ensuring that each cell carries out its specific function. Advancements in genetics and technology, such as the development of CRISPR have revolutionized our ability to edit and manipulate DNA. CRISPR is a powerful gene-editing tool that allows scientists to selectively modify specific genes within an organism's DNA sequence. This technology has the potential to cure genetic diseases, enhance crop yields, and address various environmental challenges. By editing the DNA sequence, scientists can remove harmful mutations, introduce beneficial traits, or even create entirely new genetic variations. The possibilities are virtually limitless, and the impacts on human health, agriculture, and the environment are profound. DNA plays a crucial role in the field of genetics. Its

ability to carry and transmit genetic information, undergo replication and transcription, and determine an organism's traits and
characteristics makes it a central focus of study and research.
Advancements in genetic technologies, such as CRISPR, have
opened new frontiers in gene editing and manipulation. These
breakthroughs have the potential to cure diseases, modify organisms, and shape the future of humanity. As our understanding of
DNA continues to expand, so too does our ability to unlock the
secrets of life and harness its potential for the betterment of
mankind. In recent years, the field of genetics has undergone a
remarkable transformation thanks to the development of revolutionary techniques such as gene editing and CRISPR. These
breakthroughs have the potential to not only cure diseases that
have plagued humanity for centuries but also to change the very
fabric of organisms themselves. The possibilities that these technologies present are truly staggering and have the power to reshape the future of humanity in ways we never thought possible.
One of the most exciting applications of gene editing and CRISPR
is in the field of medicine. Until now, many genetic disorders were
considered untreatable or only manageable with limited success.
With the advent of these new technologies, scientists have been
able to directly manipulate the genes responsible for these disorders, opening up a whole new world of possibilities for treatment and cure. Diseases like cystic fibrosis, sickle cell anemia,
and muscular dystrophy, which previously had no cure, are now
on the verge of being wiped out entirely. Gene editing can correct
the defective genes that cause these diseases, allowing individuals to live healthy and productive lives. This breakthrough has
the potential to revolutionize healthcare and improve the lives of
millions of people worldwide. But the impact of gene editing and

CRISPR is not limited to the field of medicine alone. These technologies also have the power to change the very nature of organisms. Scientists are now able to modify the genes of various organisms, from plants to animals, with unprecedented precision. This has far-reaching implications for agriculture, as crops can be genetically modified to enhance their nutritional content, increase yield, and become resistant to pests and diseases. This would not only solve the issue of world hunger but also reduce the need for harmful pesticides and fertilizers. Gene editing has the potential to resurrect extinct species through the modification of their genetic material, offering the tantalizing prospect of bringing back long-lost creatures and restoring fragile ecosystems. Gene editing and CRISPR have the potential to redefine our understanding of evolution itself. While traditionally, evolution has been a slow and gradual process occurring over millions of years, gene editing allows for the possibility of rapid and deliberate changes in organisms. This means that scientists can now actively guide the evolution of species, accelerating the process and potentially creating new, improved versions of organisms that are better suited to their environment. This has enormous implications for the future of humanity, as we could potentially engineer ourselves to be more resistant to diseases, improve our cognitive abilities, or even enhance our physical attributes. The possibilities are limited only by our imagination and ethical considerations. As with any new technology, gene editing and CRISPR come with their fair share of ethical concerns. The ability to manipulate the genes of organisms raises questions about the boundaries of what is morally acceptable. Should we, for instance, edit the genes of embryos to eliminate the risk of genetic diseases before birth? If we have the power to modify genes that affect traits such as intelligence or physical appearance, where do we draw the line? These are complex ethical dilemmas that must be carefully considered as we delve further into the possibilities offered by gene editing and CRISPR.

The emergence of gene editing and CRISPR has opened up a new frontier in the field of genetics. These technologies have the potential to cure diseases, reshape organisms, and even redefine our understanding of evolution itself. The implications for the future of humanity are staggering, offering immense possibilities for improvement and progress. Nonetheless, the ethical considerations surrounding this technology must be carefully addressed to ensure that we navigate these new frontiers with wisdom and responsibility. Only then will we be able to fully harness the power of gene editing and CRISPR for the betterment of humanity.

III. GENE EDITING TECHNIQUES

Gene editing techniques have revolutionized the field of genetics, offering unprecedented power to manipulate the genetic makeup of organisms. One such technique that has gained immense popularity in recent years is CRISPR-Cas9. This method allows scientists to edit genes by targeting specific DNA sequences and making precise modifications. CRISPR-Cas9 works by using a small piece of RNA called guide RNA to guide the Cas9 enzyme to the desired location in the genome. Once at the target site, Cas9 creates a double-stranded break in the DNA molecule, which triggers the cell's natural DNA repair mechanisms. Scientists can then introduce a desired DNA sequence into the cell, which will be incorporated during the repair process.

The versatility and ease of use of CRISPR-Cas9 have made it a game-changer in the field of gene editing. Its ability to target specific genes with high precision has opened up new possibilities for treating genetic disorders. In fact, researchers have already demonstrated the potential of CRISPR-Cas9 in the treatment of diseases such as Duchenne muscular dystrophy and sickle cell anemia. By correcting the underlying genetic mutations responsible for these conditions, CRISPR-Cas9 offers a promising avenue for permanent and targeted therapies.

CRISPR-Cas9 has not only revolutionized the treatment of genetic diseases but also has the potential to tackle major global health challenges. It has been successfully used to engineer mosquitoes that are resistant to malaria, a disease that affects millions of people worldwide. By introducing a gene that prevents the malaria parasite from completing its life cycle in mosquitoes, scientists hope to reduce the spread of the disease and ultimately eliminate it altogether. In addition to malaria, CRISPR-Cas9 holds promise for combating other vector-borne illnesses such as dengue fever and Zika virus. Beyond disease treatment and prevention, gene editing techniques offer a wide range of applications in various fields. In agriculture, CRISPR-Cas9 has the potential to revolutionize crop breeding by enabling the development of plants that are more resistant to pests and diseases, require less water, or have improved nutritional content. By precisely modifying the genes responsible for these traits, scientists can accelerate the breeding process and create crops that are better adapted to environmental challenges.

Gene editing techniques also hold promise in the realm of conservation biology. With the rapid decline of many species due to habitat destruction and climate change, genetic interventions may be necessary to ensure their survival. For instance, scientists are exploring the use of gene editing to combat white-nose syndrome, a fungal disease that has decimated bat populations in North America. By modifying genes related to immune responses, researchers hope to create bats that are resistant to the disease and thus prevent further population declines.

While the potential benefits of gene editing techniques are vast, they also raise ethical considerations and concerns. The ability to alter the genetic makeup of organisms opens up a Pandora's box of ethical questions, such as the potential for creating "designer babies" or enhancing human traits. The implications of gene editing on the understanding of identity, diversity, and equality are profound and require careful consideration from both the scientific and ethical perspectives. Gene editing techniques, particularly CRISPR-Cas9, have revolutionized the field of genetics and offer immense potential for the treatment of genetic diseases, the prevention of global health challenges, and the advancement of various fields such as agriculture and conservation biology. The ethical implications and concerns surrounding these techniques cannot be ignored. As we journey to the edge of science, it is crucial to carefully navigate the ethical and societal implications of gene editing in order to ensure a future that benefits all of humanity.

TRADITIONAL GENE EDITING METHODS

have played a crucial role in advancing our understanding of genetics and have paved the way for more precise and efficient techniques like CRISPR. One of the earliest methods of gene editing is called homologous recombination, which involves introducing a modified DNA sequence into a cell, allowing it to integrate into the organism's genome. While this method has been instrumental in studying gene function, it is extremely challenging and time-consuming, often resulting in low success rates. Another technique, known as zinc-finger nucleases (ZFNs), was developed in the 1990s and allowed scientists to cut and modify specific DNA sequences. ZFNs use a protein called zinc finger, which can recognize and bind to specific DNA sequences, and an enzyme called nuclease, which cuts the DNA at the targeted site. While ZFNs revolutionized the field of gene editing, they are still limited by their complex design and high cost, making them inaccessible to many researchers. Following ZFNs, another gene editing method known as transcription activator-like effector nucleases (TALENs) was developed. TALENs are similar to ZFNs but use a different type of protein, called transcription activator-like effector, to recognize specific DNA sequences. Like ZFNs, TALENs suffer from the issues of complexity and high cost, limiting their widespread adoption. Despite their limitations, these traditional gene editing methods have been instrumental in advancing the field of genetics and have set the stage for the emergence of CRISPR. Traditional gene editing methods have been invaluable for basic research purposes, allowing scientists to understand the function of genes and their role in diseases. For example, homologous recombination has been used to study the function of genes in model organisms like mice, by introducing targeted genetic mutations and observing the resulting phenotypic changes. This technique has been pivotal in uncovering the underlying mechanisms of numerous genetic disorders and has provided insights into potential therapeutic strategies. Similarly, ZFNs and TALENs have contributed significantly to our understanding of gene function, by allowing scientists to manipulate specific genes and observe the resulting effects on cellular processes. These methods have also been used to develop animal models of human diseases, providing invaluable tools for studying the pathogenesis of various disorders. The complexity and limitations of traditional gene editing methods have hindered their application in clinical settings and therapeutic interventions. The high cost and technical expertise required for designing and using ZFNs and TALENs have made them inaccessible to many researchers and healthcare professionals. These methods are limited in their precision, often resulting in off-target effects that can have detrimental consequences. This lack of precision poses significant risks in therapeutic applications, where any unintended modifications to the genome can lead to unforeseen complications. The time-consuming nature of traditional gene editing techniques makes them unsuitable for large-scale genetic modifications, which are often required in therapeutic interventions.

The emergence of CRISPR has revolutionized the field of gene editing, offering a more accessible, efficient, and precise alternative to traditional methods. CRISPR, is a naturally occurring defense mechanism found in bacteria, allowing them to defend against viral infections. Scientists have harnessed this system and repurposed it as a powerful gene editing tool. The CRISPR system consists of two main components: a guide RNA (gRNA) and an enzyme called Cas9. The gRNA is designed to recognize and bind to specific DNA sequences, guiding Cas9 to the targeted site. Once at the target, Cas9 cuts the DNA, enabling researchers to selectively modify and edit the genome.

CRISPR's simplicity and efficiency have made it a game-changer in the field of genetics. Its ease of use and low cost have democratized access to gene editing technology, allowing researchers around the world to make significant contributions to the field. CRISPR's precision and low off-target effects offer promising prospects for therapeutic applications, ranging from treating genetic disorders to developing novel cancer therapies. CRISPR has undoubtedly reshaped the future of gene editing and provided exciting new avenues for scientific exploration and medical interventions.

EXPLANATION OF CRISPR

CRISPR is a revolutionary gene editing tool that has ignited excitement and fascination in the scientific community. It is a naturally occurring system found in bacteria that allows them to defend against viruses by recognizing and destroying their DNA. In recent years, scientists have harnessed this system to develop a powerful and precise tool for editing genes in various organisms, including humans. The importance of CRISPR lies in its ability to manipulate the genetic code with incredible precision and efficiency, offering unprecedented opportunities for addressing genetic diseases, improving crop yields, and even addressing environmental challenges. Unlike other gene editing techniques, CRISPR is relatively simple, cheap, and accessible, making it an attractive option for researchers across different disciplines. The versatility of CRISPR has captured the imagination of scientists and the public alike, as it holds the potential to fundamentally reshape the future of humanity. The potential ramifications and ethical implications of CRISPR technology also raise important questions, as it opens up possibilities for altering the human germline, leading to heritable genetic changes that could be passed on to future generations. As CRISPR technology continues to evolve and advance, it is crucial that society engages in a thoughtful and inclusive dialogue to ensure its responsible and ethical use in the pursuit of a better future.

ADVANTAGES AND LIMITATIONS OF CRISPR

CRISPR-Cas9 technology, as described in the previous para-graphs, has emerged as a powerful gene editing tool, revolution-izing the field of genetics and opening up new possibilities for medical advancements. One of the key advantages of CRISPR lies in its ability to target specific genes with remarkable precision and accuracy. This has proven to be invaluable in both basic re-search and therapeutic applications. Researchers can now edit genes in a controlled and efficient manner, which not only allows for a better understanding of gene function but also holds great potential in treating genetic disorders. The ability to correct dis-ease-causing mutations is a particularly significant advantage of CRISPR. By targeting the root cause of a disease, rather than merely managing symptoms, CRISPR offers the prospect of de-veloping tailored treatments for individual patients. With further research and development, CRISPR has shown the potential to cure genetic disorders such as sickle cell disease, cystic fibrosis, and muscular dystrophy, which have previously been considered incurable. This opens up a new era in medicine, where genetic diseases may be cured rather than managed. CRISPR has demonstrated its potential in agriculture by enabling scientists to modify the genes of crops, enhancing their nutritional value, re-sistance to pests and diseases, and tolerance to environmental conditions. This has the potential to vastly improve global food security, especially in regions prone to agricultural challenges. By selectively editing plant genes, crop yields may increase, and the need for chemical pesticides may decrease, reducing the harmful impacts on the environment.

CRISPR can help address concerns related to genetically modified organisms (GMOs). With its precise targeting, CRISPR can avoid introducing foreign DNA into plants, allowing for genetic modifications that are indistinguishable from naturally occurring variations. This may alleviate some of the ethical and regulatory challenges associated with GMOs. Another advantage of CRISPR is its ease of use and affordability compared to previous gene-editing technologies. The simplicity and accessibility of the CRISPR-Cas9 system have democratized gene editing, allowing scientists around the world to utilize this tool in their research. This has significantly accelerated progress in the field and increased its potential for beneficial applications. The widespread adoption of CRISPR also fosters collaboration and knowledge sharing, as the scientific community can build upon each other's findings and work towards common goals. Despite its immense potential, CRISPR also faces several limitations and challenges that must be carefully addressed. One major concern is off-target effects, where CRISPR may introduce undesired mutations in non-targeted regions of the genome. Although considerable efforts have been made to enhance the specificity of the CRISPR system, mutation errors can still occur. This raises significant safety concerns, particularly in the context of human gene therapy, where the consequences of unintended modifications need to be thoroughly assessed and minimized.

The long-term effects and potential consequences of CRISPR gene editing are still not fully understood. Manipulating the genetic code can have unpredictable effects, and the full extent of these changes may take years or even generations to reveal themselves. This necessitates extensive research on the potential risks and ethical considerations associated with implementing CRISPR technology before widespread applications are adopted. The use of CRISPR in germ line editing, which involves modifying the DNA of sperm, eggs, or embryos, raises profound ethical questions about the potential dangers and implications of altering the human germline. Deliberate changes to the germline could have far-reaching implications, affecting not only the individuals being edited but also future generations. Consequently, there is an urgent need for a global conversation and regulations to navigate the ethical complexities of this technology.

CRISPR is a transformative gene editing tool that has the potential to advance genetic research and revolutionize various fields such as medicine and agriculture. Its advantages, including precise gene targeting, therapeutic potential, and affordability, make it highly promising for future applications. Challenges such as off-target effects, long-term consequences, and ethical considerations must be effectively addressed to harness CRISPR's full potential. Moving forward, responsible and informed use of this technology will be imperative in shaping a future where CRISPR and genetics can bring about positive changes for humanity. In recent years, the field of genetics has witnessed a revolutionary breakthrough known as CRISPR. This gene-editing technology, derived from the bacterial immune system, holds great promise for the future of humanity. With the ability to make precise changes to our DNA, CRISPR has the potential to cure diseases, modify organisms, and ultimately reshape the world as we know it. The possibilities seem boundless, as scientists delve deep into this exciting new frontier. One of the most significant implications of CRISPR lies in its potential to cure genetic disorders. Currently, many diseases are caused by faulty genes, and traditional medical treatments only address the symptoms rather than the underlying cause. With CRISPR, scientists can envision a future where these genetic disorders are eradicated. By editing the specific genes responsible for these conditions, CRISPR technology offers hope for a world free from diseases such as cystic fibrosis, Huntington's disease, and even certain types of cancer. This prospect not only holds immense value for individuals and families affected by genetic disorders but also has the potential to alleviate the burden on healthcare systems worldwide. Another intriguing application of CRISPR is its ability to modify organisms. Through genetic engineering, scientists can now manipulate the DNA of plants and animals to enhance desirable traits or reduce harmful ones. This technology presents opportunities for a more sustainable, efficient, and equitable agriculture system. For example, by modifying crops to be more resistant to pests or changing the genes of livestock to produce leaner meat, we could potentially address issues of food security and environmental sustainability. CRISPR offers hope for conservation efforts, as researchers explore the possibility of reviving extinct species or protecting endangered ones by modifying their genetic makeup. As with any powerful technology, ethical considerations must be taken into account to ensure responsible and thoughtful use. Perhaps the most awe-inspiring aspect of CRISPR is its potential to reshape the future of humanity. With the ability to edit our genes, we may no longer be limited by the constraints of natural selection. The concept of "designer babies" emerges on the horizon, where parents could potentially select specific genetic traits for their children, ranging from physical appearance to intelligence. This notion raises profound ethical questions, as it challenges our notions of identity, equality, and the natural order of life. While some argue that this kind of genetic manipulation could lead to a society divided along genetic lines, others see it as an opportunity to enhance the overall well-being of humanity. The responsible use of CRISPR in this realm requires careful consideration of the social, ethical, and legal implications. As we journey to the edge of science, the power of CRISPR beckons us to explore its vast potential. In a time where diseases continue to ravage lives and genetic disorders pose daunting challenges, gene-editing technology brings hope for a brighter future. By precisely targeting and modifying genes, CRISPR enables us to combat the root causes of illness, offering possibilities previously unimaginable. The ability to modify organisms opens the door to a more sustainable and equitable world. As with every scientific breakthrough, caution is essential. We must navigate the ethical implications of playing with the building blocks of life and temper our excitement with responsibility. Genetics and CRISPR represent a new frontier in scientific exploration with far-reaching implications for humanity. The ability to edit our genetic code holds the potential to cure diseases, modify organisms, and reshape our very nature. As scientists and society grapple with the immense power of CRISPR, it is crucial to consider the ethical implications and ensure its responsible use. By doing so, we can harness the immense power of this groundbreaking technology to usher in a future where genetic disorders are eradicated, food security is enhanced, and the human potential is maximized. The journey to the edge of science continues, and CRISPR offers a tantalizing glimpse of a future limited only by our imagination.

IV. POTENTIAL APPLICATIONS OF GENE EDITING

The potential applications of gene editing are vast and have the potential to revolutionize various fields, including healthcare, agriculture, and environmental conservation. One of the most promising areas in healthcare is the treatment of genetic diseases. Gene editing techniques, such as CRISPR, can be used to correct the underlying genetic mutations that cause diseases like cystic fibrosis, sickle cell anemia, and muscular dystrophy. By editing the genetic code, researchers hope to eradicate these diseases from future generations. Gene editing also holds the potential to revolutionize cancer treatment. By targeting and modifying cancer-causing genes, scientists aim to develop more effective and personalized therapies that can specifically target tumor cells without harming healthy ones. Another promising application of gene editing is in the field of agriculture. Traditional breeding methods to develop crops with desirable traits can be time-consuming and imprecise. With gene editing, scientists can introduce specific changes to the DNA of crops, resulting in crops that are more resistant to pests, have improved nutritional content, and can thrive in harsh environmental conditions. This has the potential to increase food production and address issues such as food security and malnutrition in developing countries. Gene editing also offers exciting possibilities in the field of environmental conservation. By modifying the genes of endangered species, scientists can potentially increase their chances of survival.

For example, researchers could edit the genes of coral species to make them more resistant to rising water temperatures, helping to protect coral reefs from the detrimental effects of climate change. Gene editing could be used to combat invasive species by modifying their genes or disrupting their reproductive abilities, which could help restore balance to ecosystems and protect native species. The potential benefits of gene editing are not limited to healthcare, agriculture, and conservation. It also has the potential to reshape the way we think about reproduction and human enhancement. Gene editing techniques like CRISPR have the potential to enable the editing of genes in human embryos, raising significant ethical questions and considerations. While it may be possible to eliminate genetic diseases and enhance certain desirable traits, such as intelligence or physical abilities, it also raises concerns about the potential for creating a genetically privileged class and the loss of genetic diversity. The concept of "designer babies" is particularly controversial, as it raises questions about the limits of human intervention in the natural process of reproduction and the potential for unintended consequences. Despite these ethical concerns, gene editing has the potential to significantly improve the lives of individuals with genetic diseases and disabilities. By enabling the correction of genetic mutations, gene editing could offer individuals the chance to live healthier and more fulfilling lives. It could also provide hope for families affected by genetic diseases, offering the possibility of eradicating these conditions from future generations.

Gene editing holds immense potential for a wide range of applications in healthcare, agriculture, environmental conservation, and human enhancement. The ability to modify the genetic code has the power to revolutionize the way we treat and prevent diseases, increase food production, protect endangered species, and potentially reshape the future of human reproduction. While the possibilities are exciting, it is crucial to proceed with caution and consider the ethical and societal implications associated with gene editing. As we embark on this new frontier of genetics, it is essential to strike a balance between scientific progress and responsible use, ensuring that the potential benefits are maximized while minimizing the risks and respecting the diverse values and perspectives of society.

GENE THERAPY FOR GENETIC DISEASES

Gene therapy offers a promising solution for the treatment of genetic diseases. By directly targeting and modifying the defective genes responsible for a particular disorder, this approach has the potential to cure diseases that were previously considered incurable. One notable example of successful gene therapy is the treatment of severe combined immunodeficiency (SCID), also known as "bubble boy" disease. SCID is a rare genetic disorder that severely compromises the immune system, leaving affected individuals vulnerable to severe infections. In the past, the only available treatment was a bone marrow transplant, which carried significant risks and was only possible if a suitable donor could be found. Thanks to gene therapy, a safer and more efficient treatment option is now available. The underlying genetic defect in SCID is a mutation in the IL2RG gene, which codes for a critical protein involved in immune system development. By using a modified virus as a delivery vehicle, scientists are able to introduce a healthy copy of the IL2RG gene into the patient's cells. These modified cells can then produce the functional protein, restoring the immune system's ability to fight infections. This groundbreaking treatment has been successfully used in clinical trials, leading to the complete recovery of several SCID patients. Another example is the treatment of certain forms of inherited blindness. In these cases, gene therapy aims to replace or supplement the malfunctioning genes that are responsible for the degeneration of retinal cells. By introducing a functional copy of the defective gene, researchers have been able to restore vision in some patients. These breakthroughs in gene therapy represent a new era in medicine, where we have the ability to directly target and correct the underlying genetic cause of diseases. While the potential of gene therapy is undeniable, there are still significant challenges that need to be addressed before it can become a widespread and accessible treatment option. One such challenge is the delivery of therapeutic genes to target cells. Currently, the most common method of delivery is the use of viruses that have been modified to carry the therapeutic gene. Although effective, this approach comes with its own set of limitations. Viruses can elicit immune responses, potentially leading to adverse reactions in patients. Viral vectors have a limited cargo capacity, which restricts the size of the therapeutic gene that can be delivered. The precise and efficient targeting of specific cells or tissues remains a challenge. Different diseases may require different cellular targets, making it crucial to develop methods that are highly specific and capable of delivering the therapeutic gene only to the intended cells. Despite these challenges, the potential of gene therapy for the treatment of genetic diseases is undeniable. As our understanding of genetics and gene editing technologies continues to advance, so too will our ability to develop safer and more efficient gene therapies. In particular, the development of CRISPR-Cas9 technology has revolutionized the field of gene editing. CRISPR-Cas9 allows scientists to precisely edit the DNA sequence of genes, providing a powerful tool for correcting genetic mutations. By using CRISPR-Cas9, researchers can delete, insert, or modify specific sections of DNA with unprecedented accuracy. This technology has the potential to transform the field of gene therapy by offering a more precise and targeted approach to gene editing. Despite its immense potential, CRISPR-Cas9 is not without its limitations. Off-target effects and unintended mutations are among the concerns that need to be carefully addressed before CRISPR-Cas9 can be safely and effectively used in a clinical setting. Nonetheless, the ability to directly edit the genetic code holds immense promise for the treatment of genetic diseases. As we continue to explore the exciting world of genetics and gene editing, we are venturing towards new frontiers that have the potential to change not only how we treat diseases but also how we perceive our own genetic makeup. The possibilities are endless, and the future of gene therapy is set to shape the future of humanity.

ALTERING CROPS TO ENHANCE AGRICULTURAL PRODUCTIVITY

Another fascinating application of gene editing technology is the ability to alter crops in order to enhance agricultural productivity. With the world's population continuing to grow, the demand for food is steadily increasing. Traditional methods of breeding and crop improvement have their limitations. They can be time-consuming, imprecise, and may not always yield the desired results. Gene editing offers a promising solution to these challenges.

One area of focus in crop enhancement is improving the nutritional content of crops. By editing the genes responsible for certain nutritional traits, scientists can create crops that are richer in vitamins, minerals, and other essential nutrients. For example, researchers have used gene editing to increase the iron content in rice, a staple crop in many developing countries where iron deficiency is prevalent. By introducing a gene from a different plant species, they were able to significantly boost the iron levels in the rice grains, potentially providing a cost-effective solution to combat malnutrition. In addition to enhancing nutritional content, gene editing can also be used to improve crop yields. Traditional breeding methods often involve crossing plants with desired traits, hoping that the offspring will inherit those traits. This process can be time-consuming and unpredictable. With gene editing, scientists can precisely target and modify specific genes responsible for traits such as disease resistance, drought tolerance, and pest resistance. By doing so, they can create crops that are more resilient and productive, potentially helping to feed the ever-growing global population. One notable example of gene editing's potential in crop improvement is the use of CRISPR technology to develop disease-resistant crops. Plant diseases are a major concern for farmers, as they can cause devastating losses in crop yield and quality. Traditional methods of disease control often involve the use of chemical pesticides, which can be harmful to the environment and human health. By using CRISPR, scientists can edit the genes of crops to make them resistant to specific diseases, eliminating the need for chemical interventions. For instance, researchers have successfully used CRISPR to create a strain of wheat that is resistant to powdery mildew, a common fungal disease that affects wheat crops worldwide. This breakthrough not only reduces the reliance on pesticides but also ensures a more sustainable and environmentally friendly approach to agriculture. Gene editing can also be used to modify crops for improved water and nutrient efficiency. As the availability of water and fertilizers becomes increasingly limited, it is crucial to develop crops that can thrive under such conditions. Through gene editing, scientists can introduce genes that enhance water and nutrient uptake, allowing crops to grow more efficiently with limited resources. This not only reduces the environmental impact of agriculture but also ensures food security in regions where water and nutrient scarcity is a pressing issue. Despite its enormous potential, gene editing in crop improvement raises ethical and regulatory concerns. The release of genetically modified organisms (GMOs) into the environment poses risks that need to be carefully considered and assessed. Questions of ownership and control over gene-edited crops and their impact on biodiversity also need to be addressed. The ethical implications of altering the genetic makeup of organisms should not be underestimated, and it is crucial to have transparent and inclusive discussions to ensure responsible and ethical use of gene editing technology in agriculture. Gene editing holds great promise for enhancing agricultural productivity and addressing global food security challenges. By precisely modifying the genes of crops, scientists can develop strains that are more nutritious, resilient, and efficient. These innovations have the potential to revolutionize agriculture and reshape the future of humanity. Careful consideration must be given to the ethical and regulatory aspects of gene editing to ensure its responsible and sustainable use. As we venture into this exciting frontier of genetics, it is vital to approach it with caution and foresight, guided by the principles of ethics and sustainability.

ENGINEERING ANIMALS FOR BETTER MEAT PRODUCTION

Engineering animals for better meat production is another area where gene editing and CRISPR technology have the potential to revolutionize the future of humanity. Meat consumption is rapidly increasing globally, and traditional methods of animal agriculture are not sustainable in the long term. By genetically modifying animals, scientists can create breeds that are more resistant to diseases, have a higher meat yield, and require less food and water. This could significantly reduce the environmental impact of meat production and provide a more efficient and environmentally friendly solution to meet the growing demand for animal products. One example of how gene editing can be utilized in animal agriculture is the development of disease-resistant livestock. Currently, livestock farmers face significant losses due to diseases such as foot-and-mouth disease, African swine fever, and avian influenza. These diseases not only pose a threat to animal health but also have the potential to spread to humans, leading to significant economic and public health consequences. By using gene editing techniques, scientists can introduce specific genes that confer resistance to these diseases into livestock breeds. This approach has proven successful in creating pigs that are resistant to African swine fever, offering hope for the future of pig farming and food security. In addition to disease resistance, gene editing can also be utilized to enhance the meat yield of livestock. Traditional breeding methods are time-consuming and often result in a loss of desirable traits during the selection process. In contrast, gene editing allows for precise modifications to specific genes, ensuring that the desired traits are retained. For example, scientists have successfully used CRISPR to create hornless dairy cows, eliminating the need for painful dehorning procedures while maintaining high milk production. This not only improves animal welfare but also reduces the risk of injury to farmers and other animals.

Gene editing can also be used to improve the feed efficiency of livestock. Livestock production is resource-intensive, requiring large amounts of feed and water. By genetically modifying animals to be more efficient in converting feed into meat, we can reduce the environmental impact of meat production. For example, scientists have successfully edited the FGF21 gene in pigs, resulting in animals that require 25% less food to produce the same amount of meat. This not only reduces the demand for feed but also decreases the production of waste and greenhouse gas emissions associated with livestock farming.

The use of gene editing in animal agriculture also raises ethical concerns. Critics argue that these technologies could be used to create animals that experience increased suffering or have compromised welfare. There is the potential for unintended consequences, as gene editing could inadvertently introduce new diseases or disrupt ecosystems. It is essential to carefully consider the ethics and potential risks associated with these technologies before implementing them on a large scale.

Gene editing and CRISPR technology offer exciting possibilities for engineering animals for better meat production. By creating disease-resistant livestock, enhancing meat yields, and improving feed efficiency, we can address many of the challenges currently facing animal agriculture. It is crucial to approach these technologies with caution and consider the ethical implications and potential risks associated with their use. As we continue to push the boundaries of genetics and explore the new frontiers of science, it is essential to strike a balance between technological advancement and the well-being of animals and the environment. The future of meat production lies in the hands of scientists, policymakers, and society as a whole, who must work together to ensure that these technologies are used responsibly for the benefit of humanity and the planet. Gene editing, particularly through the revolutionary tool CRISPR, has emerged as one of the most promising fields in genetics, offering the potential to revolutionize numerous aspects of our lives. This innovative technology allows for the precise modification of an organism's DNA, resulting in the ability to cure diseases, alter traits, and even edit the germline. The possibilities presented by CRISPR are unprecedented, holding the key to solving some of humanity's most daunting challenges. By harnessing the power of gene editing, we have the potential to cure genetic diseases that were once considered untreatable, such as cystic fibrosis and sickle cell disease. CRISPR has already shown remarkable success in laboratory settings, with experiments successfully correcting genetic mutations in living organisms. These advancements offer hope to millions of individuals and families affected by genetic disorders, providing the possibility of a future free from the limitations imposed by these conditions. Beyond disease prevention and treatment, gene editing has the potential to reshape our environment and enhance the agricultural sector. By modifying the DNA of crops and livestock, scientists using CRISPR can create organisms that are more resilient to diseases, pests, and adverse environmental conditions. This has enormous implications for food security, as it allows for the development of crops that can thrive in arid regions or withstand extreme temperatures. Gene editing can lead to a reduction in the need for harmful pesticides and herbicides, benefiting both the environment and human health. By using CRISPR to enhance the desirable qualities of crops, such as yield, taste, and nutritional content, we can create a future where global hunger is significantly reduced, and nutrition is improved for all. Perhaps the most ethically and morally complex aspect of gene editing lies in its potential to edit the germline, thus altering the genetic makeup of future generations. This new frontier raises numerous ethical questions, as it challenges our notions of what is natural and what constitutes an acceptable boundary in altering human DNA. While the prospect of eliminating inherited genetic diseases from future generations is undoubtedly enticing, it also raises concerns about the potential for creating designer babies or exacerbating existing social inequalities. These ethical considerations must be carefully evaluated and addressed for responsible and equitable implementation of gene editing technologies. The advancements in gene editing also provide exciting possibilities for combating some of the most pressing global health issues. CRISPR has the potential to eradicate mosquito-borne diseases such as malaria and dengue fever by modifying the genes of these disease-carrying insects. By creating genetically modified mosquitoes that cannot transmit these diseases, we could eliminate the devastating impact they have on human populations, particularly in developing countries. Gene editing could pave the way for the development of personalized cancer treatments by targeting specific genetic mutations present in an individual's cancer cells. The ability to tailor treatment to an individual's genetic profile holds the promise of more effective and precise therapies, improving patient outcomes and reducing the side effects associated with traditional chemotherapy and radiation treatments. It is important to acknowledge that with the unprecedented power of gene editing comes the responsibility of ethical and regulatory oversight. The potential misuse or abuse of these technologies is a concern that must be addressed to ensure that the benefits of gene editing are shared equitably and responsibly. It is crucial that these advancements are guided by rigorous scientific and ethical guidelines, with transparent discussions involving stakeholders from various fields, including science, medicine, ethics, and policy. Gene editing, particularly through the revolutionary tool CRISPR, holds immense potential to reshape the future of humanity. This groundbreaking technology offers the ability to cure diseases, enhance agricultural productivity, combat global health issues, and alter the genetic makeup of future generations. It is vital that the ethical implications of these advancements are carefully considered and that regulatory frameworks are established to ensure responsible and equitable use. As we venture into the exciting frontiers of genetics and gene editing, it is crucial to steer the development and application of these technologies towards the betterment of society, benefiting all of humanity.

V. ETHICS AND CONCERNS SURROUNDING GENE EDITING

While the potential benefits of gene editing are undeniably fascinating, there are also significant ethical and societal concerns that must be taken into consideration. First and foremost, there is the issue of consent. With the ability to manipulate an individual's genetic code, questions arise regarding the autonomy and agency of the person whose genome is being edited. Should parents have the authority to edit their child's DNA to prevent debilitating diseases or enhance desirable traits? Or does this infringe upon the child's right to make decisions about their own genetic makeup when they come of age? The concept of germline editing raises questions about the implications for future generations. By editing the genetic code of an embryo or sperm/egg cells, the changes made would be passed down to all subsequent generations, creating a permanent alteration to the human gene pool. This raises concerns about unforeseen consequences and the potential for unintended harmful effects on future populations.

Beyond issues of consent and germline editing, there are also fears of misuse and abuse of gene editing technology. As with any powerful tool, there is a risk that gene editing could be used in unethical or even harmful ways. The potential for creating "designer babies" with enhanced intelligence, athleticism, or other desirable traits raises concerns about exacerbating existing inequalities and creating a society of genetic haves and have-nots. There is the possibility of unintended consequences and a lack of understanding about the full extent of genetic interactions. Modifying one gene could inadvertently lead to negative effects on other aspects of an individual's biology, causing unforeseen health problems. Another significant ethical concern surrounding gene editing is the potential for eugenics and the erosion of diversity. With the ability to choose and manipulate desirable traits, there is a risk that certain characteristics could be deemed superior and others stigmatized or eliminated altogether. This has clear parallels to past eugenic practices that sought to improve the human population through selective breeding, and raises questions about the preservation of diversity in all its forms. Genetic diversity is essential for the resilience and adaptability of a population, and by narrowing the range of genetic traits, we may unintentionally limit our ability to respond to new challenges and threats. The idea of "playing God" and the hubris that comes with such manipulation of the fundamental building blocks of life raises philosophical and moral issues that cannot be easily overlooked. In addition to ethical concerns, there are also practical and regulatory challenges that must be addressed before widespread use of gene editing can become a reality. The technology itself is still relatively new and not yet fully understood, with many potential risks and uncertainties remaining. Ensuring the safety, efficacy, and accessibility of gene editing treatments will require stringent regulations and careful oversight. The cost and affordability of gene editing therapies may contribute to disparities in healthcare access, further exacerbating existing social inequalities. Balancing the potential benefits of gene editing with these ethical, practical, and regulatory concerns will require robust and open dialogue between scientists, policymakers, ethicists, and the broader public.

Gene editing holds immense promise for the future of medicine, agriculture, and our understanding of life itself. The ethical and societal implications must be carefully considered and addressed. Issues such as consent, germline editing, misuse, eugenics, diversity, and practical challenges all play a role in shaping the future landscape of gene editing. As we venture into this new frontier of genetics, it is imperative that we approach gene editing with caution, open-mindedness, and a commitment to ensuring the responsible and ethical use of this powerful tool. Only through careful consideration and thoughtful decision-making can we harness the potential of gene editing for the betterment of humanity while minimizing the risks and pitfalls that unbridled use may bring.

ETHICAL IMPLICATIONS OF GENE EDITING

Gene editing, particularly with the advent of CRISPR-Cas9 technology, has opened up a new frontier in the field of genetics. The potential to cure genetic diseases, change the genetic makeup of organisms, and shape the future of humanity is both exciting and daunting. The ethical implications of gene editing cannot be ignored. Gene editing raises important questions about the limits of scientific intervention, the potential for unintended consequences, and the potential for creating a divide between the genetically enhanced and the unenhanced.

One major ethical concern surrounding gene editing is the slippery slope argument. Critics argue that once we start manipulating the genetic code, it will become increasingly difficult to determine where the boundaries should lie. While gene editing may initially be used to treat genetic diseases, there is the risk that it could be taken to the extreme, resulting in the creation of "designer babies" with enhanced traits such as intelligence or physical capabilities. This raises questions about fairness and equality, as those who can afford gene editing procedures would have an advantage over those who cannot. It also brings into question the value we place on natural diversity and the potential for a loss of individuality and uniqueness. Another ethical concern is the potential for unintended consequences. Genes are interconnected and have complex interactions, so altering a single gene could have unforeseen effects on other aspects of an organism's physiology or behavior. While CRISPR-Cas9 is relatively precise, there is still a margin for error. A small mistake could lead to disastrous outcomes, such as the creation of new diseases or the disruption of ecosystems. The long-term effects of gene editing are largely unknown, making it difficult to fully assess the risks and benefits. This raises the question of whether we should proceed with gene editing without fully understanding the potential consequences. Gene editing has the potential to create a divide between the genetically enhanced and the unenhanced. If gene editing becomes widely available, it could result in a society where some individuals are genetically superior to others. This could lead to further inequalities and discrimination based on genetic traits. There is also the potential for a loss of diversity and the homogenization of traits, as individuals opt for the same desirable characteristics. This raises questions about the value of difference and the potential for a loss of acceptance and inclusivity. Questions of informed consent and autonomy arise in the context of gene editing. Should parents have the right to alter the genetic makeup of their children even before they are born? Does society have the right to dictate what traits are desirable or acceptable? These are complex questions that require careful consideration. There is the dilemma of how to regulate and enforce guidelines and restrictions surrounding gene editing. Striking a balance between promoting scientific progress and safeguarding against the potential harms and ethical pitfalls of gene editing is a challenging task. Gene editing holds immense potential for the future of medicine and human development. The ethical implications cannot be overlooked. The slippery slope argument, the potential for unintended consequences, the divide between the genetically enhanced and the unenhanced, questions of informed consent and autonomy, and the challenges of regulation and enforcement all highlight the need for careful and thoughtful discussion regarding the ethical implications of gene editing. As we navigate the new frontiers of genetics, it is crucial to consider the values, principles, and potential consequences at stake to ensure that these groundbreaking technologies are used in a manner that respects individual autonomy, promotes fairness and inclusivity, and safeguards against unintended harm. Only through this rigorous ethical examination can we truly harness the potential of gene editing for the greater good of humanity.

CONCERNS REGARDING UNINTENDED CONSEQUENCES

While the field of genetics and the development of CRISPR technology offer immense potential for advancing humanity, there are also concerns regarding unintended consequences that need to be carefully considered. One of the primary concerns surrounding gene editing is the potential for off-target effects, where unintended changes may occur in the genome of an organism. CRISPR technology relies on the use of RNA molecules to guide a protein to a specific location within the genome, where it can then make precise edits. There is still a risk that the RNA molecule may bind to an unintended location, leading to unintended changes that could have detrimental effects on the organism's health and development. Another concern is the potential for germline editing, which involves making changes to the genetic material that can be passed down to future generations. While this may offer the possibility of eradicating certain genetic diseases, it also raises ethical questions about the alteration of the human germline. The long-term consequences of these changes are still largely unknown, and there is a need for careful consideration of the implications before proceeding with germline editing. The widespread use of gene editing technology also raises concerns about equity and access. The development and application of CRISPR technology requires considerable resources, including state-of-the-art laboratories, skilled scientists, and significant financial investments. This creates a potential divide between those who have the means to use this technology for their own benefit and those who are unable to access or afford it. If gene editing becomes primarily available to the wealthy and privileged, it could exacerbate existing inequalities and further marginalize vulnerable populations. The potential misuse of this technology is a significant concern. CRISPR has the potential to be used for purposes other than curing diseases or improving health outcomes. For example, it could be used for cosmetic purposes, such as enhancing physical appearance or altering traits for non-medical reasons. This raises ethical questions about the boundaries of genetic manipulation and the potential for creating an artificial divide between those who can afford such enhancements and those who cannot. There is the risk of genetic discrimination, where individuals who have not undergone gene editing may face discrimination based on their genetic makeup. The implications of gene editing extend beyond human applications and raise concerns regarding the environment and biodiversity. The alteration of genes in organisms has the potential to create unintended ecological consequences. For example, gene editing in crops could lead to unintended effects on surrounding ecosystems, affecting pollinators, biodiversity, and potentially causing harm in unexpected ways. Careful consideration and regulation are essential to ensure that gene editing is used responsibly and does not have detrimental impacts on our environment. While the field of genetics and CRISPR technology present exciting possibilities for improving human health and shaping the future of humanity, they also come with concerns regarding unintended consequences. It is crucial to address these concerns through further research, careful regulation, and ethical considerations. By doing so, we can ensure that the potential benefits of gene editing are maximized while minimizing any potential harm or negative impacts. The use of CRISPR technology should be guided by the principles of safety, equity, and respect for the boundaries of genetic manipulation, all while considering the wider implications on our environment and future generations. Only by taking a thoughtful and balanced approach can we truly harness the potential of gene editing to enrich human lives without compromising our collective well-being.

REGULATION AND OVERSIGHT IN THE FIELD OF GENE EDITING

As the field of gene editing continues to flourish, it becomes increasingly important to establish robust regulations and oversight mechanisms to ensure its responsible and ethical implementation. The potential of gene editing technologies, such as CRISPR, to cure diseases and alter organisms is immense, but it also raises concerns about the unintended consequences and ethical dilemmas that may arise. It is crucial to strike a balance between promoting innovation and ensuring the safety and well-being of individuals and communities.

One area that requires careful regulation is the use of gene editing in human embryos. While this technology holds immense promise in terms of preventing genetic diseases and improving the health of future generations, it also raises significant ethical questions. For instance, there is the risk of unintended off-target effects and the possibility of making permanent changes to the germline, which would be passed on to future generations. The potential for designer babies, where parents can select desirable traits for their children, raises concerns about inequality and the commodification of human life. Strict regulations and oversight should be in place to govern the use of gene editing in human embryos to ensure that it is only done for medically necessary purposes and with the utmost respect for human dignity.

In addition to human embryos, the use of gene editing in non-human organisms also calls for regulatory frameworks. The modification of non-human organisms, such as plants and animals, using CRISPR has the potential to increase crop yields, develop disease-resistant animals, and restore endangered species. There are concerns about the unintended environmental impacts and the potential for genetically modified organisms to escape into the wild. To mitigate these risks, strict regulations should be put in place to ensure that gene editing in non-human organisms is conducted in a controlled and responsible manner. This includes rigorous risk assessments, environmental impact studies, and careful monitoring of any released genetically modified organisms. The commercialization of gene editing technologies necessitates robust regulatory frameworks. The market for gene editing tools and therapies is rapidly growing, with numerous companies and research institutions vying for dominance. This rush to market raises concerns about the safety and efficacy of gene editing products. There have already been instances where unregulated and untested gene therapies have caused harm to patients. In order to protect individuals and ensure the integrity of the field, regulatory agencies should establish clear guidelines for the development, testing, and approval of gene editing products. This includes rigorous clinical trials, independent oversight, and transparent reporting of results. It is crucial to strike a balance between promoting innovation and ensuring the safety and efficacy of gene editing therapies. The global nature of gene editing necessitates international cooperation and collaboration in regulation and oversight. The scientific community, regulatory agencies, and policy-makers from different countries need to work together to develop harmonized standards and guidelines. This will help prevent unethical practices, such as gene editing tourism, where individuals travel to countries with lax regulations to access gene editing procedures that are not allowed in their home country. By establishing a global framework for gene editing, we can ensure that ethical and responsible practices are adhered to across borders. As the field of gene editing progresses, it is essential to establish robust regulations and oversight mechanisms to guide its responsible and ethical implementation. Strict regulations are needed for the use of gene editing in human embryos to prevent ethical dilemmas and unsafe practices. Similar regulations are necessary for gene editing in non-human organisms to mitigate environmental risks. The commercialization of gene editing requires clear guidelines to ensure the safety and efficacy of gene editing products. International collaboration is crucial in developing harmonized standards and preventing unethical practices. By striking a balance between promoting innovation and ensuring the safety and well-being of individuals and communities, we can navigate the exciting frontiers of genetics and gene editing in a responsible and ethical manner.

Gene editing, particularly through the use of CRISPR technology, holds immense potential for revolutionizing various aspects of the future of humanity. The ability to modify genes has far-reaching implications in terms of curing diseases, altering organisms, and ultimately reshaping the very fabric of society. Firstly, gene editing has the capacity to eliminate genetic disorders that have plagued humans for generations. Diseases like cystic fibrosis, sickle cell anemia, and Huntington's disease, which have long challenged medical professionals, may soon become treatable or even curable through CRISPR. By precisely editing specific genes responsible for these conditions, it may be possible to correct the problematic genetic sequences, offering hope to the millions who suffer from these diseases. This technology has the potential to eradicate inherited disorders altogether, allowing future generations to be spared from the burden of genetic diseases.

In addition to the medical ramifications, gene editing has the capability to alter organisms in ways previously only dreamed of. For instance, CRISPR technology presents an opportunity to enhance the nutritional value of crops, ensuring food security for a growing global population. Through genetic modifications, scientists could develop crops that are more resilient to harsh weather conditions, pests, and diseases, resulting in increased crop yields and reduced dependency on chemical pesticides and fertilizers. This not only addresses the pressing issue of food scarcity but also holds the potential to mitigate the environmental impact of agriculture, benefiting both human welfare and the planet as a whole. Gene editing could be utilized to engineer livestock with desirable traits, such as disease resistance or improved meat quality, thereby enhancing animal welfare and agricultural productivity. The future of humanity could be fundamentally shaped by the potential of gene editing to influence human traits. While the notion of "designer babies" raises ethical concerns, the ability to select for desired traits holds undeniable appeal. Genetic interventions could potentially enhance intelligence, physical attributes, and even creativity, raising questions about the ethics of manipulating the natural genetic lottery. While the moral implications are complex, the potential to give future generations a head start in life cannot be ignored. Nevertheless, caution must be exercised to prevent any potential abuses of this technology, ensuring that it is used for the betterment of society as a whole rather than exacerbating disparities. Beyond the realm of biological organisms, gene editing has the potential to reshape the future of medicine itself. CRISPR can be utilized to target and alter specific genes within viruses, opening new avenues for treating viral infections. By modifying the genetic material of viruses, it may be possible to create vaccines that are more effective in preventing infectious diseases that have historically posed significant threats to humanity, such as influenza. CRISPR-based therapies may hold promise in combating antibiotic resistance, a growing global crisis. By precisely editing the genes responsible for drug resistance, scientists could potentially render previously untreatable bacterial infections vulnerable to existing antibiotics once again.

Gene editing, particularly through the use of CRISPR technology, offers exciting opportunities for advancing the future of humanity. From curing genetic diseases to enhancing crop productivity, gene editing has the potential to transform various facets of our lives. We must proceed with caution to navigate the ethical and moral questions that arise from this groundbreaking technology. By treading carefully and considering the long-term implications, we can harness the power of gene editing to shape a future that improves the lives of individuals while benefiting society as a whole. The journey to the edge of science has never been more thrilling, and the potential that lies within gene editing has the capacity to unlock new frontiers for the betterment of humanity.

VI. DISEASE TREATMENT AND PREVENTION

In addition to revolutionizing agriculture and bioengineering, gene editing technologies such as CRISPR have the potential to transform the field of medicine by providing innovative approaches to disease treatment and prevention. One area where these advancements are particularly promising is with regards to genetic disorders. In the past, patients suffering from genetic diseases had limited treatment options, often relying on palliative care to manage symptoms. With the advent of CRISPR, researchers now have the ability to directly target and modify the underlying genetic mutations responsible for these disorders. By using CRISPR to edit the faulty genes, it becomes possible to potentially cure these diseases at their root cause. This groundbreaking approach has already shown tremendous potential in preclinical studies, with researchers successfully using CRISPR to correct genetic mutations in cells and animal models of diseases such as sickle cell anemia, cystic fibrosis, and Duchenne muscular dystrophy. CRISPR-based gene editing techniques have the potential to enhance the effectiveness of traditional drug therapies. One challenge in drug development is ensuring that medications reach their intended targets within the body without causing harm to healthy cells. Gene editing technologies, including CRISPR, offer a potential solution to this problem. By delivering therapeutic genes directly into the patient's own cells using CRISPR, researchers can potentially bypass the need for traditional drug delivery methods altogether. This targeted approach allows for higher concentrations of therapeutic agents to be delivered precisely to the diseased cells, effectively increasing the efficacy of the treatment while minimizing side effects. CRISPR can be used to modify the patient's own cells to make them more receptive to specific medications, improving the overall response to drug therapies. In the realm of infectious diseases, CRISPR has the potential to revolutionize disease prevention strategies. Traditional approaches to combatting infectious diseases often involve the development of vaccines or the use of antibiotics. These methods are sometimes inadequate due to the ability of pathogens to evolve and develop resistance. CRISPR offers a new approach to combatting infectious diseases by enabling researchers to directly interfere with the genetic material of pathogens. By targeting and editing specific genes in the pathogen's genome, CRISPR can potentially disrupt essential biological processes or render the pathogen unable to evade the immune system. This technology has already demonstrated promise in battling infections such as human immunodeficiency virus (HIV), malaria, and antibiotic-resistant strains of bacteria. CRISPR can play a crucial role in disease surveillance and early detection. One of the challenges in the field of diagnostic medicine is the rapid identification of pathogens in order to initiate appropriate treatment. Current methods often require time-consuming laboratory testing and analysis. CRISPR-based diagnostic tools, known as CRISPR-based detection assays, offer a rapid and specific means of detecting and identifying disease-causing agents. These assays work by leveraging the CRISPR system's ability to bind and cleave specific target sequences of DNA or RNA. This targeted approach allows for the quick identification of specific pathogens or genetic markers associated with disease. These novel diagnostic tools have the potential to significantly enhance disease surveillance efforts, enabling healthcare professionals to identify outbreaks and respond swiftly to emerging infectious diseases. The advancements in gene editing technologies such as CRISPR have immense potential to transform disease treatment and prevention strategies. From curing genetic disorders and enhancing traditional drug therapies to revolutionizing infectious disease prevention and diagnostics, CRISPR offers a new frontier in medicine. As researchers continue to explore and refine the capabilities of these breakthrough technologies, the future of humanity holds the promise of improved health outcomes and a new era in the fight against diseases. The exciting journey to the edge of science has only just begun, and the implications for humanity are vast and profound.

CRISPR'S POTENTIAL TO CURE GENETIC DISEASES

CRISPR has emerged as a revolutionary tool in the field of genetics. This groundbreaking technology, coupled with the newfound ability to edit genes, holds immense potential for finding cures for genetic diseases. CRISPR allows scientists to precisely cut and modify DNA strands, providing a way to correct the genetic mutations responsible for diseases such as cystic fibrosis, sickle cell anemia, and Huntington's disease. By targeting the specific genes causing these ailments, CRISPR offers the hope of permanently alleviating suffering for millions of individuals worldwide. One of the more promising areas where CRISPR has shown potential is in treating sickle cell anemia. This inherited blood disorder affects millions of people globally, predominantly those of African descent. The disease is caused by a single mutation in the beta-globin gene, leading to the formation of abnormal hemoglobin molecules that result in misshapen red blood cells. These deformed cells often get stuck in blood vessels, causing pain, organ damage, and a shorter lifespan. Current treatments for sickle cell anemia are limited to managing symptoms and preventing complications, leaving patients with a reduced quality of life. With the advent of CRISPR, researchers have successfully used this gene-editing tool to correct the genetic mutation responsible for sickle cell anemia in laboratory experiments. By introducing a modified copy of the beta-globin gene into the patient's stem cells, CRISPR has the potential to eradicate the disease at its root, providing a long-lasting and perhaps even permanent cure.

Another genetic disease that CRISPR holds promise for treating is cystic fibrosis. This life-threatening disorder affects the lungs, pancreas, and other organs, leading to the accumulation of thick, sticky mucus. This mucus clogs the airways, making it difficult to breathe, and creates a breeding ground for infections. Cystic fibrosis is caused by mutations in the CFTR gene, which encodes a protein responsible for transporting chloride ions across cell membranes. These mutations disrupt the normal functioning of the CFTR protein, resulting in the thickened mucus seen in individuals with cystic fibrosis. Like sickle cell anemia, current treatments for cystic fibrosis aim to manage symptoms and slow disease progression. CRISPR offers the potential to correct the defective CFTR gene, restoring normal protein function and reversing the effects of the disease. This approach could potentially provide a cure for cystic fibrosis, improving the quality of life for affected individuals and increasing their life expectancy. CRISPR has the potential to revolutionize gene therapy by offering a way to treat genetic diseases that were previously deemed untreatable. Conditions caused by a single gene mutation, such as Huntington's disease, are prime candidates for the application of CRISPR. Huntington's disease is a degenerative neurological disorder that leads to the progressive breakdown of nerve cells in the brain, resulting in severe cognitive and motor impairments. It is caused by an abnormal expansion of a trinucleotide repeat in the HTT gene, leading to the production of a toxic protein that damages neurons. While there is currently no cure for Huntington's disease, CRISPR provides a glimmer of hope for those affected. By using CRISPR to selectively edit and remove the expanded trinucleotide repeat, researchers have successfully corrected the genetic mutation in experimental models of the disease. Such breakthroughs hold promise for the development of a potential cure for Huntington's disease, offering relief to those suffering from this debilitating condition.

CRISPR's potential to cure genetic diseases is nothing short of remarkable. This powerful gene-editing tool has opened up a new frontier in genetics, offering hope for finding cures for previously untreatable conditions. By precisely cutting and modifying DNA, CRISPR allows scientists to correct the genetic mutations responsible for diseases like sickle cell anemia, cystic fibrosis, and Huntington's disease. The success of CRISPR in laboratory experiments has paved the way for future clinical applications, with the potential to permanently alleviate the suffering of millions of individuals worldwide. While ethical and safety concerns remain, the promise that CRISPR holds for curing genetic diseases is undoubtedly exciting and represents a significant leap forward in the field of genetics. With continued advancements and research, CRISPR may reshape the future of humanity, offering the possibility of a world free from the burden of genetic diseases.

ELIMINATING INFECTIOUS DISEASES USING GENE EDITING

One of the most promising applications of gene editing is the possibility of eliminating infectious diseases. Infectious diseases have been a persistent threat to humanity since time immemorial, causing immense suffering and death. With the advent of gene editing technologies such as CRISPR, we now have the tools to potentially eradicate these diseases once and for all.

Traditionally, infectious diseases have been treated with antibiotics or vaccines. While these interventions have been successful in many cases, they are not without their limitations. Antibiotics, for instance, can become ineffective due to the emergence of drug-resistant bacteria, a phenomenon that has become a major public health concern in recent years. The development of vaccines can be time-consuming and costly, making it difficult to respond quickly to emerging infectious threats.

Gene editing offers a new approach to combating infectious diseases. By directly targeting the genetic material of the disease-causing organisms, gene editing can effectively render them harmless. This approach has already shown great promise in the fight against malaria, a disease that claims hundreds of thousands of lives every year. Researchers have successfully used gene editing techniques to modify mosquitoes, the primary vectors for the malaria parasite, in a way that can prevent them from transmitting the disease to humans. By introducing genetic modifications that make the mosquitoes resistant to the parasite, scientists have been able to significantly reduce the incidence of malaria in controlled laboratory settings. If these genetically modified mosquitoes were released into the wild, they could potentially help to eradicate malaria altogether.

In addition to malaria, gene editing holds great potential for eliminating other infectious diseases as well. For instance, HIV/AIDS, a global health crisis that affects millions of people, could potentially be eradicated using gene editing. Researchers have made significant progress in developing gene editing techniques that can remove the HIV virus from infected cells, effectively curing patients of the disease. While there are still many technical and ethical challenges to overcome before this approach can become a reality, it represents a promising avenue for a disease that has defied conventional therapies for decades. Gene editing could also be used to eliminate other viral diseases such as influenza or hepatitis. By targeting specific genes in the virus, scientists could potentially disrupt the virus's ability to replicate or infect host cells, effectively rendering it harmless. This approach could not only lead to the development of new treatments for these diseases but could also help to prevent future outbreaks or pandemics. Gene editing could be used to engineer livestock and crops that are resistant to diseases, thereby improving food security and reducing the risk of infectious diseases spreading to humans. Despite its immense potential, gene editing for the purpose of eliminating infectious diseases also raises important ethical considerations. The ability to manipulate the genetic material of organisms raises questions about the boundaries of scientific intervention and the potential consequences of altering natural ecosystems. There are also concerns about the potential misuse of gene editing technologies, such as the creation of genetically modified organisms for nefarious purposes.

These ethical considerations must be carefully addressed and regulated to ensure that gene editing is used responsibly and for the benefit of humanity. Gene editing holds great promise for eliminating infectious diseases and reshaping the future of humanity. By directly targeting the genetic material of disease-causing organisms, gene editing techniques such as CRISPR offer a new approach to combating infectious diseases that is potentially more effective and efficient than traditional methods. From malaria to HIV/AIDS, gene editing has the potential to eradicate some of the world's most devastating diseases. The ethical implications of gene editing must also be carefully considered and regulated to ensure that these technologies are used responsibly. As we venture into the new frontiers of genetics, the possibilities for improving human health and well-being are truly remarkable.

PREVENTATIVE GENE EDITING FOR DISEASE SUSCEPTIBILITY

Preventative gene editing holds the potential to revolutionize medicine by targeting disease susceptibility at its source. With the advent of gene editing technologies such as CRISPR, scientists now have the ability to make precise modifications to the genetic code, effectively altering an individual's DNA. This raises the possibility of not only treating diseases after they occur, but also preventing them from ever developing in the first place. By editing certain genes that are known to increase one's susceptibility to diseases, scientists can potentially eliminate the risk entirely. This has enormous implications for both individuals and society as a whole. One of the key advantages of preventative gene editing is the ability to tackle genetic diseases. Many diseases, such as cystic fibrosis and Huntington's disease, are caused by mutations in specific genes. By using CRISPR to edit these disease-causing genes, scientists can potentially eliminate the genetic basis of these diseases altogether. This would have a profound impact on individuals who are at risk of developing such diseases, as well as their families. No longer would they have to live in fear of inheriting a debilitating condition, as preventative gene editing could ensure that the disease-causing genes are removed from their genetic makeup. This could even be done at the embryonic stage, allowing individuals to be born completely disease-free. In addition to genetic diseases, preventative gene editing could also target common diseases that have a genetic component. Many diseases, such as cancer and heart disease, have a complex interplay between genetic and environmental factors. While it may not be possible to completely eliminate the risk of these diseases through gene editing alone, it could significantly reduce an individual's susceptibility. By editing genes that are known to increase the risk of these diseases, scientists could potentially lower the chances of individuals developing them. This would not only have a significant impact on the individuals themselves, but also on society as a whole. With the prevalence of these diseases on the rise, preventative gene editing could potentially save millions of lives and reduce the burden on healthcare systems worldwide.

Preventative gene editing could also address the issue of antibiotic resistance. Antibiotic resistance is a growing global concern, with many bacteria becoming increasingly immune to the effects of antibiotics. This poses a significant threat to public health, as diseases that were once easily treatable could become untreatable in the future. By using gene editing technologies, scientists could potentially modify the genetic code of bacteria to make them more susceptible to antibiotics. By targeting specific genes involved in antibiotic resistance, scientists could potentially reverse this trend and make antibiotics effective once again. This has the potential to save countless lives and mitigate the future impact of antibiotic resistance. While preventative gene editing holds immense promise, it also raises ethical and societal concerns. The ability to alter an individual's genetic code raises questions about the limits of scientific intervention and the potential for unintended consequences. There is also the issue of access and equity, as gene editing technologies could potentially be limited to those who can afford them, further exacerbating existing health disparities. There are concerns about the potential for genetic enhancement rather than just disease prevention. This could lead to a widening gap between genetically enhanced individuals and those who are not, potentially creating a new form of inequality. Preventative gene editing has the potential to be a groundbreaking advancement in medicine, allowing for the prevention of genetic diseases, reducing susceptibility to common diseases, and addressing antibiotic resistance. It also raises ethical and societal concerns that must be carefully considered. As we venture into the new frontiers of genetics, it is crucial that we approach these technologies with caution and consider both the potential benefits and risks they pose. The future of humanity lies in our ability to navigate these challenges and make informed decisions about the responsible use of gene editing technologies. This is a journey that will shape the future of medicine and reshape the very fabric of our existence.

Despite the rapid advancement of genetics and gene editing technologies, there are potential ethical dilemmas and concerns associated with the use of CRISPR. One such concern is the possibility of unintended consequences and off-target effects. In the process of modifying genes, there is a risk of unintentionally altering other genes or causing unexpected mutations that can have detrimental effects on an organism. This is particularly worrisome when it comes to human gene editing, as any unintended consequences could have far-reaching implications for future generations. There are concerns about the potential for misuse or abuse of these technologies. With the ability to edit genes, there arises the temptation to use these tools for non-medical purposes, such as enhancing physical traits and intelligence. This raises questions about the ethics of creating "designer babies" and exacerbating existing disparities between individuals. The accessibility and affordability of CRISPR technologies could potentially lead to a divide between those who can afford gene editing and those who cannot, creating a new form of inequality in society. The lack of long-term studies on the effects of gene editing raises concerns about the safety and potential risks associated with these technologies. While there have been significant breakthroughs in gene editing, there is still much that is not fully understood about the intricacies of genetics, and the long-term effects of manipulating genes are still largely unknown. This lack of knowledge poses a risk when it comes to implementing these technologies on a large scale. There are also societal and cultural implications to consider in the field of genetics and gene editing. Different cultures and religions may have varying beliefs and values when it comes to manipulating genetic material. These ethical concerns need to be carefully addressed and respected in order to ensure that the future of genetics and gene editing is inclusive and respects the diverse perspectives of different communities. Despite these challenges and concerns, the potential benefits of genetics and gene editing are vast and cannot be ignored. The ability to cure genetic diseases, eradicate harmful mutations, and enhance human health and well-being holds immense promise. With the development of these technologies, we have the potential to eradicate devastating illnesses, such as cystic fibrosis and Huntington's disease, improving the lives of countless individuals and families. Gene editing technologies could offer solutions to global challenges such as food security and environmental conservation. By modifying crops to be more resilient, nutritious, and resistant to pests, we could address issues such as malnutrition and the depletion of natural resources. Similarly, the ability to modify organisms to clean up pollution or degrade harmful substances could have a significant impact on environmental sustainability. The frontier of genetics and CRISPR holds great promise for the future of humanity. It is a realm where diseases can be cured, organisms can be changed, and the possibilities for reshaping our world seem boundless. It is crucial that we approach these advancements with careful consideration and a commitment to ethical principles. By exercising responsible decision-making, promoting inclusivity, and addressing concerns regarding safety and accessibility, we can harness the full potential of gene editing technologies for the betterment of humanity. As we embark on this journey to the edge of science, we must remain vigilant and uphold the integrity of genetics research, ensuring that these breakthroughs benefit society as a whole and pave the way for a brighter future.

VII. ENHANCING HUMAN ABILITIES AND TRAITS

With the incredible advancements in genetics and the advent of CRISPR technology, scientists have now entered the realm of enhancing human abilities and traits. This scientific frontier not only promises to cure previously untreatable diseases but also provides the potential to enhance and modify certain human characteristics. While this concept raises ethical concerns, the possibilities it offers for the future of humanity cannot be ignored.

One area where genetic enhancement holds tremendous potential is in the treatment and prevention of genetic disorders. CRISPR has revolutionized the field of gene editing by allowing scientists to precisely target and modify specific genes. With this technology, diseases caused by a single faulty gene, such as sickle cell anemia, Huntington's disease, or cystic fibrosis, could potentially be cured at their root cause. By editing the genes responsible for these conditions, scientists can effectively eliminate the risk of passing on the disease to future generations. This not only offers hope to countless individuals and families affected by genetic disorders but also has the potential to eradicate these conditions altogether. In addition to curing diseases, genetic enhancement also holds promise in improving human abilities. It is conceivable that in the not-so-distant future, we may be able to enhance our physical and cognitive abilities through genetic modifications. For example, athletes could potentially have their genes edited to increase muscle strength or endurance, pushing the limits of human performance. Similarly, individuals with genetic predispositions to certain mental illnesses could have their genes modified to reduce the risk of developing these conditions. By altering the genetic code, we could potentially unlock abilities and traits that were previously inaccessible, opening up new possibilities for human potential. The concept of genetic enhancement also raises ethical concerns and prompts important questions about the boundaries of science and the implications of altering the human genome. Critics argue that genetic enhancement could perpetuate existing inequalities and create a divide between a genetically enhanced elite and the rest of society. They argue that genetic enhancements, if easily accessible to only a privileged few, could exacerbate existing social, economic, and health disparities. Concerns about unintended consequences and unforeseen side effects of genetic modifications cannot be dismissed. The long-term effects of genetic alterations on individuals and future generations are still largely unknown, and therefore, proceeding with caution is crucial.

Despite the ethical concerns, the potential benefits of genetic enhancement cannot be overlooked. The ability to cure diseases, enhance abilities, and reshape the future of humanity is a journey that science cannot afford to ignore. With proper regulations and safeguards in place, genetic enhancement could become a tool for progress, improving the quality of life for individuals and society as a whole. To ensure the responsible use of genetic enhancements, it is imperative that strict ethical guidelines and regulations be put in place. These guidelines must encompass issues such as access to genetic enhancements, potential exploitation, and the long-term effects on individuals and the wider society. Public awareness and education on the science of gene editing and its potential impact are crucial to fostering informed discussions and decision-making. Only through an informed and ethical approach can we navigate this new scientific frontier and harness its potential for the betterment of humanity.

The advancements in genetics and CRISPR technology open up a world of possibilities in enhancing human abilities and traits. The potential to cure diseases, improve physical and cognitive abilities, and reshape the future of humanity is both exciting and daunting. This exciting frontier also raises ethical concerns and prompts important questions about the implications of altering the human genome. To navigate this new field responsibly, it is essential to establish strict regulations, educate the public, and ensure that the benefits of genetic enhancements are accessible to all while minimizing potential risks. The future of genetic enhancement holds incredible potential, and it is up to us to shape it responsibly.

GENE EDITING TO ENHANCE COGNITIVE ABILITIES

Gene editing has the potential to revolutionize humanity by enhancing cognitive abilities. The ability to manipulate our genetic code opens up new frontiers in neurobiology and neuroscience. By precisely editing genes that affect cognitive function, scientists may be able to enhance intelligence, memory, and other mental capacities. This has profound implications for various aspects of society, including education, employment, and even the potential for creating a new breed of super-intelligent individuals. Gene editing technologies like CRISPR have the potential to correct genetic mutations that cause cognitive disorders such as Alzheimer's disease and autism, providing hope for millions of affected individuals and their families. With the ability to edit genes directly, scientists can now target the underlying genetic causes of these disorders, potentially leading to effective treatments and even prevention. As with any powerful technology, gene editing to enhance cognitive abilities raises ethical concerns. The concept of genetically enhancing intelligence has been met with both excitement and apprehension. On one hand, it offers the potential for significant advancements in human capabilities and the opportunity to alleviate the burden of cognitive disorders. On the other hand, it raises ethical questions about inequality, discrimination, and the potential for creating a "genetic elite" that could exacerbate social and economic disparities. Critics argue that gene editing to enhance cognitive abilities could exacerbate existing inequalities and create a world where certain individuals have an unfair advantage over others. There are concerns about the unintended consequences of gene editing, as altering one gene could have unforeseen effects on other aspects of an individual's biology or even on future generations. Despite these ethical concerns, the field of gene editing to enhance cognitive abilities continues to advance rapidly. Researchers are constantly uncovering new genes and genetic pathways that influence cognitive function, bringing us closer to understanding the complex interplay between genes and the brain. CRISPR technology is becoming faster, cheaper, and more precise, making gene editing increasingly feasible and accessible. As gene editing technologies continue to evolve, it is essential that we grapple with the ethical questions they raise. Society must engage in thoughtful and ongoing dialogue regarding the potential risks and benefits of gene editing for cognitive enhancement, weighing the desire for human improvement against the potential for unintended consequences and the perpetuation of inequality. Regulatory frameworks and ethical guidelines must be established to ensure that gene editing is used responsibly and equitably. Gene editing to enhance cognitive abilities holds immense promise, but it also demands careful consideration and responsible implementation. As our understanding of genetics and the human brain deepens, we must confront the ethical dilemmas posed by this new frontier of science. With open and informed discussion, we can navigate the complex challenges and shape a future where gene editing advances the well-being of all humanity.

DESIGNER BABIES AND ETHICAL CONSIDERATIONS

The potential to genetically modify embryos and create "designer babies" raises a host of ethical considerations. On one hand, proponents argue that genetic engineering could be used to eliminate genetic diseases and create healthier individuals. By identifying and removing harmful genetic mutations, parents could ensure that their children have the best chance at a long and fulfilling life. Genetic modification could potentially enhance traits such as intelligence or athleticism, opening up possibilities for individuals to have desirable characteristics that were previously out of reach. This utopian vision of a world with disease-free and genetically superior individuals is certainly alluring. There are also significant ethical concerns associated with this technology. First and foremost, there are concerns about equity and access. If genetic engineering becomes widely available, it is likely that only those with sufficient financial resources will be able to afford it. This could exacerbate existing social inequalities, as individuals born into less affluent families would be at a genetic disadvantage. This raises questions about the fairness and justice of allowing some individuals to have access to enhanced genetic traits while others do not. There are concerns about the potential for misuse and abuse of genetic engineering. Once the technology becomes available, it is difficult to predict how it will be used. Will it be used solely for medical purposes, or will individuals and societies start to use it for more superficial reasons, such as selecting for specific physical features or enhancing cosmetic traits? The line between medical necessity and personal preference can be blurry, and there is a real risk that this technology could be exploited for non-medical purposes.

Another ethical concern is the potential for unintended consequences. The long-term effects of genetic modification are largely unknown, and there is the possibility that unintended mutations or unforeseen health risks could arise. This raises questions about the responsibility of scientists and society at large to ensure the safety and well-being of future generations. Is it ethical to experiment on embryos when we cannot be certain of the potential risks and consequences? There are concerns about the impact of genetic engineering on human diversity and the concept of what it means to be human. By allowing individuals to select and manipulate genetic traits, we risk homogenizing the population and erasing the natural diversity that exists. Genetic diversity is essential for the survival and adaptation of the human species, and eliminating or controlling it could have unforeseen consequences. It raises questions about the concept of autonomy and individuality. If we start to view genetic traits as something that can be chosen and controlled, what does this mean for personal identity and the idea of being a unique individual? There are concerns about the implications of genetic engineering for future generations. Once genetic modifications are made, they are heritable and will be passed on to future generations. This raises questions about consent and the rights of unborn children. By making choices about their genetic makeup, are we infringing upon their autonomy and right to self-determination? This also raises concerns about the potential for creating a genetic underclass or caste system, where those who have not been genetically modified are viewed as less than desirable.

The potential to create "designer babies" through genetic engineering raises a host of ethical considerations. While the idea of eliminating genetic diseases and enhancing desirable traits is enticing, there are significant concerns about equity, misuse, unintended consequences, impact on diversity, and the rights of future generations. As we navigate the potential of genetic engineering, it is essential that we engage in thoughtful and ethical considerations to ensure that we are shaping a future that is fair, just, and respectful of human rights and values. Only by doing so can we truly harness the power of genetics to improve the human condition without sacrificing our fundamental ethical principles.

IMPACT ON SOCIETAL NORMS AND EQUALITY

The developments in the field of genetics and gene editing, particularly with the advent of CRISPR technology, have the potential to revolutionize societal norms and greatly impact equality. One significant aspect of this potential impact lies in the realm of disease prevention and treatment. By understanding and manipulating our genetic code, scientists can potentially eradicate or prevent the inheritance of genetic diseases. This would not only alleviate suffering for individuals and families affected by such conditions but would also lead to a shift in societal norms. The stigma and discrimination associated with genetic diseases could gradually disappear as these conditions become rare or non-existent, opening doors for a society that values inclusivity and compassion. CRISPR technology holds the promise of addressing certain ethical concerns related to reproduction. Currently, individuals who carry certain genetic diseases face difficult decisions when considering having children, as they risk passing on the condition to their offspring. With gene editing techniques, it would be possible to eliminate or correct the anomaly in the embryo, thereby effectively preventing the disease from being inherited in the first place. This would alleviate the burden of parents having to make difficult choices while promoting the freedom of individuals to start families without fear or uncertainty. In addition to disease prevention, gene editing has the potential to expand the boundaries of human capabilities. Genetic modifications could enhance certain traits such as intelligence, strength, or physical appearance. While this may initially seem like a positive advancement, concerns about the creation of a genetic hierarchy and exacerbation of existing social inequalities arise. If only the wealthy and privileged have access to genetic enhancements, it could lead to a deepening of societal divisions, where a genetically enhanced elite emerges. This potential scenario raises questions about fairness, justice, and the nature of equality in a society where individuals are no longer on a level playing field due to their genetic makeup.

The impact of gene editing on gender equality is another area worth considering. Currently, gender inequality persists in various ways, including disparities in wages, opportunities, and societal expectations. It is conceivable that gene editing could be used to modify embryos to ensure the gender of a child, potentially perpetuating existing gender imbalances. This could reinforce harmful stereotypes and further entrench gender inequality in society. It is crucial that ethical considerations and regulations are put in place to prevent the misuse of gene editing technology in ways that exacerbate existing inequalities. The potential for unintended consequences cannot be overlooked. In the pursuit of genetic manipulation, there is a chance that unforeseen genetic mutations or long-term effects could emerge. This raises concerns about the implications for human health and well-being, as well as potential impacts on future generations. Policies and regulations must be in place to ensure that rigorous testing and assessment are conducted to minimize risks and protect the well-being of individuals and society as a whole.

Overall, the developments in genetics and gene editing, particularly with the advent of CRISPR, hold immense promise for the future of humanity. From the prevention and treatment of genetic diseases to addressing ethical concerns related to reproduction, gene editing has the potential to reshape societal norms. It is essential to approach this technology with caution, considering its potential impacts on societal equality and ethical dilemmas. By carefully considering the implications and implementing comprehensive regulations, society can harness the power of genetic advancements in a way that promotes inclusivity, justice, and the betterment of humanity as a whole. This exciting field of genetics offers us the opportunity to transcend our limitations, but it is up to us to ensure that these advancements promote social progress rather than perpetuate inequality and division.

One of the most exciting frontiers in the field of genetics is the advent of gene editing, specifically the revolutionary technology known as CRISPR. With the ability to act as molecular scissors, CRISPR has the potential to cure diseases, change organisms, and reshape the future of humanity. This exciting world of genetics holds great promise for the field of medicine and has already yielded remarkable results in the treatment of various genetic disorders. By harnessing the power of CRISPR, scientists can edit the DNA of living organisms with unparalleled precision, correcting mutations that cause genetic diseases.

This remarkable technology has the potential to transform the lives of millions of people who currently suffer from genetic conditions, offering hope for a future free from the burden of such diseases. CRISPR has the ability to change organisms in ways that were once unimaginable. By targeting specific genes, scientists can modify traits that are inherited, ultimately altering the genetic makeup of an entire species. This has far-reaching implications, particularly in the realm of agriculture, as it could potentially eliminate diseases that decimate crops or increase the nutritional value of staple foods. CRISPR has the potential to target and eliminate harmful organisms, such as disease-causing mosquitoes, thus drastically reducing the spread of deadly diseases such as malaria and dengue fever. These applications can have a profound impact on the health and well-being of individuals and communities around the globe. With such immense potential, it is no wonder that the world of genetics and CRISPR is seen as the new frontier of science. In the quest to unlock the secrets of genetics, scientists have embarked on a journey to the edge of science. This journey involves not only the exploration of the limitless possibilities of gene editing, but also the ethical and moral implications that arise from wielding such power. While the potential benefits of CRISPR are undeniable, there are concerns about the potential misuse of this technology. Gene editing could be used to enhance desirable traits, creating a society where only the genetically enhanced have access to opportunities and advantages. This brings to light questions of fairness, equality, and the potential for discrimination. Altering the genetic makeup of organisms could have unintended consequences on ecosystems, disrupting delicate balances and potentially causing unforeseen harm. These ethical and environmental considerations highlight the need for careful regulation and oversight. As humanity ventures into these new frontiers of genetics, it is crucial to address these ethical dilemmas and ensure that the benefits of this technology are accessible to all, while mitigating its potential risks.

Despite the ethical and regulatory challenges, the future of genetics and CRISPR is undeniably exciting. As our understanding of the human genome grows, the possibilities for gene editing become increasingly vast. Scientists are already exploring the potential of CRISPR to treat complex diseases such as cancer and HIV, offering hope for revolutionary breakthroughs in the field of medicine. As our knowledge of genetics expands, we may be able to unravel the mysteries of aging, potentially leading to interventions that extend human lifespan and improve the quality of life for individuals around the world. The implications of such advancements are profound, shaping the very fabric of humanity and redefining our relationship with the natural world.

The world of genetics and CRISPR holds immense potential for the future of humanity. The revolutionary technology of CRISPR allows scientists to cure diseases, change organisms, and reshape the world around us with unprecedented precision. From the treatment of genetic disorders to the modification of inherited traits, the possibilities are boundless. As we embark upon this thrilling journey to the edge of science, it is crucial to address the ethical and environmental implications of gene editing and ensure that this technology is used responsibly. With careful regulation and oversight, the future of genetics and CRISPR can be a bright and promising one, offering hope for a world where genetic diseases are a thing of the past and the potential of human existence can be unlocked to its fullest extent.

VIII. ENVIRONMENTAL IMPLICATIONS

While the potential of gene editing and CRISPR technology to cure diseases and enhance human abilities is thrilling, it also raises serious concerns regarding its environmental implications. As we venture into the new frontiers of genetics, we must consider the potential consequences of manipulating the genetic makeup of organisms and releasing them into the environment.

One of the primary worries is the unintended ecological impacts that genetically modified organisms (GMOs) may have. Altering the genetic traits of plants and animals can lead to unforeseen consequences within ecosystems. For instance, if a genetically modified crop is introduced into the wild, it can crossbreed with wild relatives, resulting in the spread of modified genes throughout the plant population. This could potentially lead to the displacement of native species, disrupt ecosystems, and cause a decline in biodiversity. The release of gene-edited organisms into the environment poses the risk of unintended consequences that could be irreversible. We cannot predict the long-term effects of genetically engineered organisms on the delicate balance of ecosystems. These modified organisms may have unforeseen behaviors, adaptations, or interactions with other species that could have detrimental effects on the environment. Once released, it may be challenging, if not impossible, to control or reverse these ecological impacts. Gene editing techniques also raise concerns regarding biosecurity and the potential for unintended consequences in relation to agriculture. The development of gene drives, for instance, threatens to spread genes intentionally

131

throughout populations, affecting entire ecosystems. While gene drives have the potential to control and eradicate harmful pests, such as disease-carrying mosquitoes, the unintended consequences of releasing such organisms into the environment must be carefully considered. There is the possibility that these gene drives could spread to unintended targets, leading to the extinction of non-target species or ecological imbalances.

In addition to the risks associated with gene editing technologies, the use of CRISPR also presents challenges related to ethical and societal implications. The ability to modify the genetic makeup of organisms raises questions about whether we should play the role of "master" over nature and intervene in the natural order of things. The potential for creating designer babies and altering human traits poses ethical dilemmas that must be carefully evaluated. The commercialization and patenting of gene editing technologies raise concerns about access and equity. Will gene editing and CRISPR be accessible only to the affluent, further exacerbating existing inequalities? The cost of these technologies and the potential for commercial interests to control access raise important social and economic questions.

To mitigate the potential environmental risks and address the ethical and societal considerations, robust regulatory frameworks and international collaborations are essential. Governments, scientists, and policy-makers must work together to establish guidelines and regulations that ensure rigorous safety assessments and prevent the unintended consequences of gene editing technologies. This includes evaluating the potential impacts on ecosystems and biodiversity, as well as addressing issues of biosecurity and equity. Public engagement and education are crucial. As society grapples with the ethical and social implications of gene editing and CRISPR, it is essential to involve the public in discussions and decision-making processes. Public awareness campaigns and educational initiatives must be implemented to ensure that the risks and benefits of these technologies are understood by all. This will foster a more inclusive and informed dialogue surrounding the future of genetic engineering and its impact on humanity and the environment.

The emerging field of gene editing and CRISPR technology has the potential to revolutionize medicine, agriculture, and conservation efforts. We must proceed with caution and consider the environmental, biosecurity, ethical, and social implications that come along with these breakthroughs. Responsible research, robust regulation, and public engagement are crucial to ensure that we harness the power of genetics for the betterment of humanity and the preservation of our environment. The journey to the edge of science is undoubtedly exciting, but we must navigate it with careful consideration.

USING GENE EDITING TO COMBAT CLIMATE CHANGE

One of the most innovative applications of gene editing lies in its potential to combat climate change. As the Earth's climate continues to shift at an alarming rate, scientists are seeking new ways to mitigate the effects of global warming and create a more sustainable future. Gene editing offers a promising solution by enabling us to modify the genetic makeup of organisms to adapt to changing environmental conditions and reduce their carbon footprint. By altering the genes of plants and animals, we can create drought-resistant crops, breed heat-tolerant livestock, and engineer carbon-absorbing trees. These genetically modified organisms have the potential to not only survive in harsh climates but also to provide sustainable sources of food, mitigate soil erosion, and sequester carbon dioxide.

One example of using gene editing to combat climate change is the development of drought-resistant crops. With the changing climate and increasing water scarcity, traditional crops struggle to survive in arid conditions. Scientists have identified specific genes responsible for drought tolerance in certain plants and have successfully introduced these genes into other crop species using gene editing techniques such as CRISPR. As a result, these genetically modified crops are better equipped to withstand prolonged periods of drought, ensuring a stable supply of food in regions prone to extreme weather conditions.

These drought-resistant crops require less water to grow, reducing water consumption in agriculture, a sector responsible for a significant portion of global water consumption. This technological advancement not only addresses food security but also contributes to the conservation of limited water resources, a critical component of adapting to climate change.

In addition to modifying plant genes, gene editing can also be applied to livestock to create heat-tolerant breeds. As global temperatures continue to rise, livestock such as cattle and poultry face increased heat stress, resulting in reduced productivity and even death. By using gene editing techniques, scientists can identify and modify genes related to heat tolerance in animals. For example, geneticists have introduced the gene responsible for heat tolerance in camels into cattle, enabling them to withstand high temperatures. With these genetically modified animals, livestock farming can continue in regions with extreme heat, ensuring a stable supply of meat and dairy products even in a changing climate. By reducing heat stress in animals, gene editing not only improves animal welfare but also contributes to food security by sustaining livestock production in the face of rising temperatures. Gene editing can be utilized to engineer trees that have enhanced carbon-absorbing capabilities. Trees play a crucial role in combatting climate change by sequestering carbon dioxide through photosynthesis. Not all tree species are equally efficient in capturing carbon, and some are even more susceptible to the adverse effects of climate change. By identifying and modifying the genes associated with carbon uptake, scientists can create tree species that are better at absorbing and storing carbon dioxide. These genetically modified trees could be strategically planted in areas with high carbon emissions, such as near factories or urban centers, to mitigate the impact of greenhouse gases. These trees can also contribute to land restoration efforts by displacing invasive species and preventing soil erosion. By using gene editing to enhance the carbon-absorbing capabilities of trees, we can effectively reduce the levels of atmospheric carbon dioxide, one of the primary drivers of global warming.

The application of gene editing in combating climate change offers immense potential to create a more sustainable future for humanity. By using gene editing techniques such as CRISPR, we can modify the genetic makeup of organisms to adapt to changing environmental conditions. From developing drought-resistant crops and heat-tolerant livestock to engineering carbon-absorbing trees, gene editing provides innovative solutions to mitigate the effects of global warming. These genetically modified organisms not only provide sustainable sources of food but also contribute to water conservation, enhance animal welfare, and reduce greenhouse gas emissions. As we continue to explore the possibilities and ethical considerations of gene editing, its application in addressing climate change offers a glimpse into the exciting future of genetics and the new frontiers of science.

MODIFYING ORGANISMS TO ADAPT TO HARSH CONDITIONS

Modifying organisms to adapt to harsh conditions is an area of research that holds great promise for the future. With the advancement of genetic engineering and CRISPR technology, scientists have been able to make significant strides in modifying the genetic makeup of organisms to enhance their abilities to thrive in extreme environments. This emerging field of research raises intriguing possibilities for a variety of applications, from agriculture to space exploration. In the field of agriculture, modifying organisms to adapt to harsh conditions has the potential to revolutionize farming practices and address the growing global food crisis. By genetically engineering crops to withstand drought, high temperatures, or salinity, scientists can ensure that food production remains stable even in regions with unfavorable environmental conditions. This could help alleviate hunger and malnutrition in many parts of the world, where arid or infertile soils have traditionally hindered agricultural productivity. Genetically modified crops could reduce the need for harmful pesticides and fertilizers, promoting a more sustainable and environmentally friendly approach to farming. In addition to addressing global food security, modifying organisms to adapt to harsh conditions could also have a profound impact on human health. Through genetic engineering, scientists can potentially modify human cells to better withstand certain diseases or conditions. For instance, individuals with genetic disorders such as sickle cell anemia or cystic fibrosis could benefit from gene editing techniques that modify their own DNA to correct the underlying genetic mutations. In the future, this technology could potentially extend to modifying embryos to prevent hereditary diseases from being passed on to future generations.

Another area where modifying organisms to adapt to harsh conditions is showing great promise is in space exploration. As humans venture out into the cosmos, they will be faced with environments that are vastly different from Earth. The ability to modify organisms to survive in extreme conditions, such as low gravity or high levels of radiation, could be critical for long-duration space missions. By engineering organisms to produce their own oxygen or provide sustenance without the need for external supplies, astronauts could become more self-sufficient during their journeys. Genetically modified organisms could play a role in terraforming other planets, making them habitable for human colonization. The path to modifying organisms to adapt to harsh conditions is not without its ethical and practical challenges. The potential for unintended consequences or unforeseen ecological disruptions cannot be overlooked. Any modifications made to organisms must be carefully considered and thoroughly tested to ensure their safety and potential benefits outweigh the risks. The social and cultural aspects of genetic modification need to be addressed, as societal acceptance and regulation of these technologies will be crucial for their responsible and equitable use.

The ability to modify organisms to adapt to harsh conditions opens up new frontiers in the field of genetics. From agriculture to space exploration, this technology holds the potential to revolutionize numerous industries and reshape the future of humanity. By enhancing the resilience of crops, addressing genetic disorders, and enabling space colonization, scientists are pushing the boundaries of what is possible in the realm of genetic engineering. Careful consideration and proper regulation must accompany these advancements to ensure their responsible and ethical application. The journey to the edge of science is paved with both promise and responsibility, and the future holds great potential for the field of modifying organisms to adapt to harsh conditions.

THE ECOLOGICAL BALANCE AND POTENTIAL RISKS OF GENETIC MODIFICATIONS

While the potential of genetic modifications and gene editing technologies such as CRISPR to cure diseases and reshape the future of humanity is indeed exciting, it is important to consider the potential risks and ecological implications that these advancements may bring. One major concern is the disruption of the delicate ecological balance that exists within natural ecosystems. Genetic modifications have the potential to introduce foreign genetic material into an organism, altering its characteristics and potentially giving it an advantage over other species. This poses a significant risk to biodiversity as these genetically modified organisms could outcompete and displace native species, leading to a loss of genetic diversity and potentially destabilizing entire ecosystems. The use of genetic modifications in agriculture, such as the creation of genetically modified crops, also raises concerns. While these crops may have desirable traits such as increased resistance to pests or enhanced nutritional value, there is a potential for unintended consequences. For example, the widespread use of genetically modified crops could lead to the evolution of resistant pests, resulting in the need for even stronger pesticides. This not only poses risks to human health and the environment but also raises questions about the sustainability of relying on genetic modifications as a solution to agricultural challenges. Another significant concern is the potential for unintended and unpredictable effects of genetic modifications. While scientists have made impressive strides in understanding and controlling gene editing technologies, there is still much that is unknown. The complex interactions between genes and their environment make it difficult to accurately predict the long-term effects of genetic modifications. This raises ethical questions surrounding the release of genetically modified organisms into the environment and the potential for unintended harm to both human health and the ecosystem.

The introduction of genetically modified organisms into the environment also raises questions of ownership and control. Who has the right to make decisions about the release of genetically modified organisms into the wild? And who is responsible for monitoring the potential impacts and mitigating any negative effects? These questions become even more complex when considering the potential for "gene drives" a technology that allows for the rapid spread of desirable genetic modifications through wild populations. While gene drives have the potential to combat diseases such as malaria and invasive species, they also raise concerns about the potential for unintended and irreversible ecological changes. In order to navigate the potential risks and ecological implications of genetic modifications, it is crucial to prioritize sound scientific research and rigorous regulation. Robust risk assessment protocols must be in place to evaluate the potential impacts of genetic modifications before they are released into the environment. This involves not only evaluating the potential risks but also carefully considering the ethical implications and public perception of these advancements.

Transparency and public engagement are essential in ensuring informed decision-making and fostering trust in the scientific community. It is important to create spaces where experts, policymakers, and the public can come together to discuss the potential risks and benefits of genetic modifications, as well as the societal and ethical implications. This will allow for a more holistic and democratic approach in navigating the complex challenges posed by genetic modifications.

While the breakthroughs in genetics and CRISPR hold immense potential to improve human health and transform the future of humanity, it is crucial to approach these advancements with caution and accountability. The ecological balance and potential risks associated with genetic modifications must be carefully considered and regulated. By prioritizing sound science, ethical decision-making, and transparency, we can harness the power of genetic modifications while ensuring the long-term sustainability and well-being of our planet. One of the most exciting developments in the field of genetics is the advent of gene editing technologies, particularly CRISPR. This groundbreaking tool, CRISPR, has the potential to revolutionize the way we approach disease treatment, reshape organisms, and even alter the future of humanity itself. With CRISPR, researchers can precisely modify the DNA of living organisms, allowing them to edit genetic information with unprecedented accuracy and efficiency. This ability holds immense promise for the cure and prevention of genetic diseases, as it allows for the targeted correction of underlying genetic mutations. By editing the DNA of affected individuals, scientists can potentially eliminate disease-causing genes or repair faulty ones, offering hope to millions of people suffering from genetic disorders. CRISPR also enables the alteration of specific traits in organisms, opening up possibilities for the creation of genetically modified crops that are more resistant to disease or produce higher yields. Gene editing can be employed to combat the spread of diseases carried by insects, such as malaria or Zika virus, by engineering mosquitoes that are unable to transmit these diseases. The applications of CRISPR extend beyond the realm of disease treatment and agriculture. It has the potential to reshape the future of humanity by offering us the ability to modify our own genetic makeup. This raises profound ethical questions and concerns about the potential consequences of such interventions. While some argue that we should embrace the possibilities of genetic enhancement and modification, touting the potential for preventing disease and enhancing human capabilities, others fear the long-term consequences and potential for misuse. The prospect of creating "designer babies" who possess specific desired traits such as intelligence or physical characteristics raises concerns about the potential for creating an "elite" class, leaving those without access to these technologies at a disadvantage. Modifications made to the germline, or reproductive cells, could be passed on to future generations, resulting in a permanent alteration of the human gene pool. The ethical and societal implications of gene editing technologies like CRISPR thus warrant careful consideration and informed public debate. Despite the ethical concerns, the potential benefits of gene editing technologies such as CRISPR cannot be overlooked. The ability to eliminate genetic diseases and enhance traits associated with a higher quality of life offers hope for countless individuals and families affected by debilitating genetic disorders. The field of gene editing has the potential to revolutionize medical research and drug development. By creating animal models with specific genetic modifications, scientists can gain a better understanding of the underlying mechanisms of disease, leading to the development of more effective treatments. CRISPR can be used to develop more precise and efficient methods of gene therapy, which holds great promise for the treatment and potential cure of a wide range of diseases, including cancer, HIV, and genetic disorders. Despite the tremendous accomplishments that have been made with gene editing technologies, there are still numerous challenges that need to be addressed before they can be widely implemented. One such challenge is the issue of off-target effects, where unintended genetic modifications occur in unrelated areas of the genome. This can have unintended consequences, potentially leading to the development of new diseases or harmful side effects. There are concerns about the safety and potential long-term effects of germline modifications. The field of gene editing is still in its infancy, and further research is needed to fully understand these potential risks and develop strategies to mitigate them. Nevertheless, the exciting potential of gene editing technologies like CRISPR cannot be denied. They offer unprecedented opportunities to reshape the future of medicine, agriculture, and even humanity itself. The ethical and societal implications of harnessing these capabilities must be carefully considered. As we venture into the new frontiers of genetics, it is imperative that we navigate this uncharted territory with caution and deliberation, ensuring that the benefits of gene editing technologies outweigh the potential risks and that the future they shape is one that we can embrace with confidence.

IX. AGRICULTURAL REVOLUTION

The potential impact of genetic engineering is not limited to the medical field alone, as it has the ability to revolutionize agriculture as well. The agricultural revolution, like the medical revolution, has the potential to address some of the most pressing challenges facing humanity today. With a rapidly growing global population and dwindling resources, the need for efficient and sustainable agricultural practices has never been greater. Genetic engineering offers a promising solution to these challenges by providing the means to enhance crop yield, increase resistance to pests and diseases, and even develop novel food sources.

One of the most significant advances in agricultural genetic engineering is the creation of genetically modified (GM) crops. These crops have been genetically modified to exhibit traits such as increased yield, improved nutritional content, and resistance to pests or herbicides. For example, Bt cotton, a variety of cotton genetically modified to produce a toxin from the bacterium Bacillus thuringiensis, is highly resistant to cotton bollworm, a common cotton pest. This not only reduces the need for chemical pesticides but also increases the yield of cotton, benefiting farmers and consumers alike. In addition to improving crop yield and resistance, genetic engineering can also enhance the nutritional content of crops, thus addressing malnutrition and nutrient deficiencies prevalent in many parts of the world. Golden Rice is a prime example of such an endeavor. It is a genetically modified variety of rice that has been modified to produce beta-carotene, a precursor of vitamin A. Vitamin A deficiency is a leading cause of blindness and childhood mortality in developing countries, and Golden Rice offers a potential solution to this global health problem. By increasing the nutritional value of staple crops, genetic engineering has the power to improve the overall health and well-being of millions of people around the world.

Genetic engineering can be employed to develop crops that are more resistant to environmental stressors, such as drought and salinity, thus increasing the resilience of agricultural systems in the face of climate change. By introducing genes responsible for enhanced water use efficiency or salt tolerance, scientists can create crops that are better able to survive and thrive in challenging environmental conditions. This is particularly relevant in regions experiencing water scarcity or prone to soil degradation, where traditional crops may struggle to grow. By harnessing the power of genetic engineering, we can develop crops that are adapted to adverse conditions, ensuring food security for future generations. Genetic engineering offers the possibility of developing novel food sources that are more sustainable and environmentally friendly than traditional agriculture. For instance, scientists are exploring the use of genetically modified microorganisms to produce alternatives to conventional meat. By engineering microorganisms to produce proteins with meat-like properties, we could potentially reduce the environmental impact associated with livestock farming, such as greenhouse gas emissions and land use. Genetic engineering can enable the production of crops with enhanced oil content, which can be utilized to produce biofuels, thus reducing our dependence on fossil fuels. These innovative approaches hold great promise for addressing the dual challenges of food security and environmental sustainability.

While the potential benefits of genetic engineering in agriculture are undeniable, it is important to consider the ethical and environmental implications associated with these technologies. The release of genetically modified organisms into the environment can have unintended consequences, such as the spread of engineered genes to wild populations and the emergence of resistant pests or weeds. The concentration of genetic modification in a few crop varieties may lead to a loss of agricultural biodiversity, increasing the vulnerability of our agricultural systems to future threats. It is essential that genetic engineering in agriculture is accompanied by rigorous risk assessment and regulatory oversight to ensure its safe and responsible application.

The agricultural revolution brought about by genetic engineering has the potential to address some of the most pressing challenges facing humanity today. Through the development of genetically modified crops, we can enhance crop yield, improve nutritional content, and increase resistance to pests and environmental stressors. Genetic engineering enables the creation of novel and sustainable food sources. It is imperative that we proceed with caution and consider the ethical and environmental implications associated with these technologies. By doing so, we can harness the power of genetic engineering to build a more sustainable and food-secure future for all.

GENE EDITING FOR CROP IMPROVEMENT

Gene editing holds immense potential for crop improvement and has already begun to revolutionize agriculture. By utilizing CRISPR-Cas9, scientists can precisely alter the DNA of crops, resulting in enhanced traits such as disease resistance, increased nutritional value, and improved yields. One prime example of this is the development of powdery mildew-resistant wheat, achieved through the introduction of a gene from a wild relative. This breakthrough not only offers a sustainable solution to combat a common wheat disease but also reduces the need for chemical pesticides, making agriculture more environmentally friendly. Gene editing techniques can be employed to address global food security challenges. Scientists have successfully edited a gene in maize, leading to a 50% increase in its yield. This remarkable achievement has the potential to greatly alleviate hunger and malnutrition around the world. Gene editing can also be used to tackle the effects of climate change on crops. By modifying genes responsible for drought tolerance or heat resistance, crops can be engineered to withstand harsh environmental conditions, ensuring stable food production in the face of rising temperatures and erratic weather patterns. Gene editing for crop improvement not only exhibits promising outcomes but also presents ethical considerations. Critics argue that the manipulation of DNA raises concerns about the unintended consequences on ecosystems and biodiversity. The possible escape of genetically modified organisms into the wild might have irreversible effects on natural ecosystems. The patenting of gene edited crops by agribusiness corporations could exacerbate global inequality and hinder small-scale farmers' access to these technologies. Hence, it is crucial to carefully evaluate and regulate gene editing practices to ensure their responsible and equitable application in agriculture. Public acceptance and informed dialogue are essential to address the concerns and fears associated with genetic manipulation and gene editing. Public engagement initiatives should aim to educate people about the benefits, potential risks, and ethical implications of gene editing for crop improvement.

Establishing transparent regulatory frameworks that involve input from various stakeholders can help safeguard against potential misuse or unintended consequences of these technologies. Despite the challenges and ethical considerations, gene editing for crop improvement represents a breakthrough in the field of agriculture that holds immense promise. By harnessing the power of gene editing, we have the ability to address pressing issues such as crop diseases, food insecurity, and the impacts of climate change on agricultural productivity. The potential to enhance crop traits precisely and efficiently can pave the way for sustainable and resilient farming systems. Gene editing techniques can contribute to reducing the environmental footprint of agriculture, through targeted improvements in crop traits that minimize the need for chemical inputs and enhance resource use efficiency. Gene editing for crop improvement can play a pivotal role in ensuring global food security and sustainable agricultural development. As we venture into the new frontiers of genetics, it is essential to embrace responsible and ethical practices that prioritize the well-being of both humans and the environment. With rigorous scientific research, transparent regulatory frameworks, and inclusive public engagement, we can harness the power of gene editing to shape a future where agriculture is not only productive but also sustainable and equitable. The possibilities offered by gene editing are indeed exciting, and as the journey to the edge of science continues, we must remain vigilant in our pursuit of a future that balances innovation with careful consideration and responsibility.

CREATING DISEASE-RESISTANT PLANTS THROUGH GENETIC ENGINEERING

Creating disease-resistant plants through genetic engineering is a promising area of research that has the potential to revolutionize agriculture and food security. As the world's population continues to grow, there is an increasing demand for more sustainable and resilient crop varieties that can withstand environmental stresses, pests, and diseases. Genetic engineering offers a powerful tool to achieve this by introducing specific genes into plants that confer resistance to various pathogens. One example is the development of genetically modified (GM) crops such as Bt corn, which expresses a toxin derived from the bacteria Bacillus thuringiensis to protect against insect pests. By incorporating these genes into the plants' DNA, scientists can effectively eliminate the need for harmful chemical pesticides, making agriculture more environmentally friendly and reducing the health risks associated with pesticide exposure. Genetic engineering can be used to enhance plants' immune systems, making them more resistant to diseases caused by bacteria, fungi, and viruses. For instance, scientists have successfully engineered resistance genes into crops like potatoes, tomatoes, and bananas, providing them with enhanced protection against devastating pathogens that can cause significant crop losses. This approach has the potential to greatly improve global food security by reducing yield losses and ensuring a steady supply of nutritious crops. Genetic engineering can help combat plant diseases that are difficult to control through conventional breeding methods. For instance, citrus greening, a bacterial disease that affects citrus trees, has caused widespread devastation in many parts of the world, leading to economic losses in the billions of dollars. By using genetic engineering techniques, scientists have been able to develop genetically modified citrus trees that are resistant to this disease. These disease-resistant trees not only offer a solution to the citrus industry but also serve as a model for other crops facing similar challenges. Genetic engineering can also help address the emerging threats posed by climate change. As the global climate continues to change, farmers are faced with increasingly unpredictable weather patterns, including drought, heatwaves, and extreme cold. These environmental stresses can have detrimental effects on crop productivity and yield. By using genetic engineering, scientists can introduce genes that confer tolerance to these stresses, allowing plants to thrive under adverse conditions. For example, researchers have successfully engineered rice with improved drought tolerance, enabling it to survive in regions prone to water scarcity. Similarly, genetic engineering has been used to develop crops that can withstand extreme temperatures, ensuring a stable food supply in the face of climate uncertainty. Creating disease-resistant plants through genetic engineering holds tremendous promise for improving agriculture and food security. By introducing genes that confer resistance to pathogens and environmental stresses, scientists can develop crop varieties that are more resilient and productive. This technology not only reduces the reliance on chemical pesticides but also helps combat emerging plant diseases and mitigate the impacts of climate change. It is important to proceed with caution and ensure thorough risk assessments are conducted to address any potential ecological or health concerns associated with genetically modified crops. Nonetheless, with continued advancements in genetics and CRISPR technologies, the future of disease-resistant plants looks bright, offering new opportunities to transform agriculture and ensure a sustainable future for humanity.

THE IMPACT ON GLOBAL FOOD SECURITY AND AGRICULTURAL PRACTICES

Gene editing and CRISPR technology have the potential to revolutionize global food security and agricultural practices. These breakthroughs offer promising solutions to the pressing challenges faced by farmers and the growing population worldwide. By improving plant traits such as yield, disease resistance, and nutritional content, gene editing can enhance crop productivity and reduce the reliance on chemical pesticides and fertilizers.

One of the primary concerns for global food security is the need to produce more food to feed the growing population. With the world's population projected to reach 9.7 billion by 2050, traditional breeding methods alone may not be sufficient to meet this demand. Gene editing and CRISPR technology open up new possibilities for increasing crop yields without the need for extensive land expansion. Through targeted modifications, scientists can enhance the genetic traits responsible for plant growth and productivity, leading to higher crop yields per unit of land. This increase in productivity can play a vital role in ensuring food security, especially in regions where agricultural resources are limited. In addition to increasing crop yields, gene editing can also address issues related to plant diseases and pests. Traditional breeding techniques often take a long time to develop disease-resistant varieties, making it challenging to respond to emerging threats effectively. The overuse of chemical pesticides and fertilizers has led to environmental pollution and adverse effects on human health. Gene editing offers a precise and efficient method to introduce disease resistance traits into crops, minimizing the need for chemical interventions. By editing specific genes responsible for disease susceptibility, researchers can develop plants that are naturally resistant to pathogens, reducing the risk of crop losses and minimizing the environmental impact of chemical treatments. Gene editing can improve the nutritional content of crops, offering potential solutions to address malnutrition and nutrient deficiencies. Many staple crops lack important nutrients necessary for human health, leading to widespread malnutrition in developing countries. By precisely modifying the genes responsible for nutrient content, gene editing can enhance the nutritional value of crops, ensuring a more balanced diet for populations dependent on these staples. For instance, scientists have used CRISPR-Cas9 to edit the genes in rice to enhance its vitamin A content, addressing vitamin A deficiency prevalent in many developing countries. These advancements in crop biofortification can have a significant impact on global health and nutrition, particularly for vulnerable populations who rely heavily on specific crops for sustenance. Despite the immense potential of gene editing in agriculture, there are ethical and regulatory considerations that must be taken into account. The consequences of unintended changes or off-target effects resulting from gene editing require careful assessment, particularly with regards to long-term impacts on biodiversity and ecosystem stability. The potential for misuse of gene editing technology raises concerns about genetically modified organisms (GMOs) and the consumer acceptance of these products. The accessibility and affordability of gene editing technology, particularly in resource-constrained regions, need to be addressed to ensure the equitable distribution of benefits across the globe. Gene editing and CRISPR technology have the potential to revolutionize global food security and agricultural practices. By increasing crop yields, enhancing disease resistance, and improving nutritional content, these breakthroughs offer promising solutions to the challenges faced by farmers and the growing population. Ethical, regulatory, and accessibility considerations must be carefully addressed to ensure the responsible and equitable deployment of this technology. Gene editing has the power to reshape the future of agriculture, offering sustainable and efficient solutions to feed the world's population while minimizing the environmental impact of conventional agricultural practices. This exciting field represents a new frontier that can unlock the potential for a more secure and sustainable future for humanity. Gene editing and CRISPR technology have revolutionized the world of genetics, opening up new possibilities and frontiers that were once thought to be mere flights of imagination. With their potential to cure diseases, change organisms, and reshape the future of humanity, these breakthroughs have propelled us into an exciting new era of scientific exploration. At the core of this revolution lies the remarkable power of CRISPR-Cas9, a gene-editing tool that enables scientists to precisely modify DNA sequences. This groundbreaking technology holds the promise of eradicating genetic defects and diseases by correcting faulty genes at their source. By targeting specific sections of DNA, CRISPR allows for the manipulation and fine-tuning of genetic information, leading to the possibility of preventing conditions such as cystic fibrosis, Huntington's disease, and even some types of cancer. Such prospects are nothing short of awe-inspiring, as they pave the way for a future in which previously untreatable genetic disorders may become curable, transforming the lives of millions around the world. Yet, CRISPR's potential extends far beyond the realm of human health. This powerful tool has the capacity to reshape the natural world, enabling scientists to modify the genes of animals and plants. By precisely altering DNA sequences, researchers hope to create more resilient and productive crops, capable of withstanding climate change and feeding an ever-growing global population. It opens up the possibility of genetically engineering animals to yield valuable resources, or even eradicating invasive species that threaten ecosystems. The implications for our environment and food security are immense, offering a glimmer of hope in the face of mounting global challenges. It is not just the practical applications of gene editing that have captured the imagination of scientists and the public alike. CRISPR has sparked ethical and moral debates, raising profound questions about the boundaries of human interference in the natural order of life. While the potential benefits of gene editing are vast, there are also concerns that tinkering with the fundamental building blocks of life may have unforeseen consequences. The ability to modify the genetic makeup of organisms opens up the doors to deliberate genetic enhancements, sparking fears of a world of designer babies and genetic inequality. This dark side of gene editing raises questions about the ethics of altering the human genome for non-medical purposes, and whether such practices would lead to a world divided between those who can afford genetic enhancements and those who cannot. In light of these concerns, ethical frameworks and robust regulations are needed to ensure that gene editing is used responsibly and for the greater good of humanity. As we venture further into the uncharted territory of gene editing and CRISPR technology, we must also address the impact they have on our understanding of what it means to be human. The ability to alter our genetic blueprint challenges our perception of nature and questions the limits of our agency over our own existence. This prompts us to ponder the age-old philosophical questions of identity and free will, as we grapple with the notion that our genetic makeup may shape who we are to a greater extent than we previously thought. At the same time, the advent of gene editing raises the possibility of rewriting the very code of life itself, bringing us closer than ever to the realm of science fiction. It presents a tantalizing vision of a future in which we can control our evolution, enhancing our physical and cognitive abilities and creating a new breed of humans. This vision comes with its own set of ethical and existential dilemmas, as we navigate the uncertain path of altering our own species. The frontiers of genetics and CRISPR technology offer a tantalizing glimpse into a future that is both exhilarating and fraught with ethical complexities. The potential to cure diseases, reshape our environment, and redefine what it means to be human hangs in the balance. As we continue to push the boundaries of scientific knowledge, it is imperative that we approach gene editing with caution, balancing the promises of progress with the need for responsible and ethical research. The journey to the edge of science beckons, and it is up to us to ensure that we navigate this newfound frontier with wisdom, foresight, and a commitment to the well-being of all.

X. THE SOCIO-ECONOMIC DIVIDE

The socio-economic divide poses significant challenges to the equitable distribution of benefits and access to the new frontiers of genetics, such as gene editing and CRISPR technology. While these breakthroughs hold immense promise for curing diseases, changing organisms, and reshaping the future of humanity, their potential impact may not be realized by all segments of society. The high cost of genetic therapies and technologies often makes them inaccessible to individuals from lower socio-economic backgrounds. This creates a divide between those who can afford such treatments and those who cannot, resulting in a further exacerbation of existing inequities in healthcare. The lack of awareness and education about these advancements in genetics can also contribute to the socio-economic divide. Individuals from privileged backgrounds are more likely to have access to information, resources, and opportunities that allow them to stay connected with the latest scientific developments. On the other hand, individuals from marginalized communities may lack access to the same resources and opportunities, limiting their ability to benefit from and contribute to the field of genetics. As a result, the socio-economic divide serves as a barrier to the full democratization of genetic advancements, raising questions about the ethical implications of unequal access to potentially life-saving treatments and interventions. It also highlights the urgent need for policies and initiatives that aim to bridge this gap and ensure the equitable distribution of benefits. By addressing the socio-economic factors that perpetuate the divide, it may be possible to create a more inclusive and accessible genetic landscape that benefits all of humanity, regardless of their socio-economic circumstances. The socio-economic divide also has implications for the future of genetic research and innovation. The lack of diversity among researchers and scientists from different socio-economic backgrounds can limit the perspectives and insights brought to the field. This homogeneity in the scientific community may lead to a skewed focus on certain genetic diseases and interventions that are more relevant to individuals from higher socio-economic groups. Consequently, the needs and concerns of marginalized communities may be overlooked or underrepresented in genetic research and its applications. To fully harness the potential of genetics and CRISPR in improving human health and well-being, it is crucial to foster diversity and inclusivity in the scientific community. By promoting equal opportunities for individuals from all socio-economic backgrounds to pursue careers in genetics and related fields, we can ensure that diverse perspectives are incorporated into research and decision-making processes. This will help in identifying and addressing the unique genetic challenges faced by marginalized communities, thus enabling more tailored and effective interventions. The socio-economic divide also intersects with other factors such as race, ethnicity, and gender, further complicating the issue. Marginalized groups often face multiple forms of discrimination that limit their access to education, healthcare, and employment opportunities, exacerbating the socio-economic divide. Recognizing and addressing these interlocking systems of oppression is essential to promoting equitable access to the new frontiers of genetics. The socio-economic divide presents a significant challenge to the equitable distribution of benefits and access to the exciting world of genetics and CRISPR technology. The high costs and lack of awareness about genetic advancements create barriers that disproportionately affect individuals from lower socio-economic backgrounds. To bridge this divide, it is crucial to implement policies and initiatives that aim to democratize genetic advancements and promote inclusivity in the scientific community. By addressing the socio-economic factors that perpetuate inequities, we can create a more equitable and accessible genetic landscape that benefits all of humanity. Fostering diversity and inclusivity in genetic research is crucial to ensuring that diverse perspectives are incorporated into decision-making processes and identifying the unique genetic challenges faced by marginalized communities. Addressing the socio-economic divide is not only an ethical imperative but also essential for harnessing the full potential of genetics and CRISPR technology in improving human health and well-being.

ACCESSIBILITY AND AFFORDABILITY OF GENE EDITING TECHNOLOGIES

The accessibility and affordability of gene editing technologies, such as CRISPR, is a topic of concern and opportunity. While these new advancements in genetics have the potential to revolutionize healthcare and biotechnology, there are also ethical and social implications that must be addressed. On one hand, the increased accessibility and affordability of gene editing technologies can democratize access to healthcare and lead to significant advancements in the treatment and prevention of genetic diseases. Currently, the cost of gene editing technologies can be prohibitively expensive, making them inaccessible to many individuals and communities. As technology improves and becomes more widely available, the costs are expected to decrease. This opens up the possibility of wider adoption and use of gene editing technologies, allowing for greater equity in healthcare. Affordability can facilitate the utilization of gene editing technologies in developing countries, where genetic diseases are prevalent but resources are limited. The potential impact on public health is significant, as these technologies have the potential to eradicate genetic diseases and significantly improve the quality and longevity of life for countless individuals. On the other hand, the accessibility and affordability of gene editing technologies raise important ethical concerns. With the potential to alter the human genome, questions of consent, privacy, and the potential for unintended consequences arise. It is essential that rigorous regulatory frameworks are established to ensure that these technologies are used responsibly and with utmost regard for human rights and dignity. The implications of gene editing technologies also extend beyond healthcare. They have the potential to reshape the future of humanity in ways that were once unimaginable. This presents both exciting opportunities and daunting challenges. By altering the genetic makeup of organisms, gene editing technologies can potentially create new species that are more resistant to disease, more resilient to environmental changes, and more efficient in various industries such as agriculture and energy production. For example, gene editing could be used to create crops that are resistant to drought or pests, thereby increasing yields and improving food security. Similarly, it could be used to engineer microorganisms that can produce biofuels, reducing reliance on fossil fuels and mitigating climate change. These advancements also raise concerns about unintended consequences and potential risks to ecosystems. The long-term effects of genetically modified organisms on the environment are not fully understood, and it is crucial to conduct thorough research and risk assessments. The potential to alter the human genome raises moral and ethical questions about what it means to be human and the boundaries of scientific intervention. Will gene editing technologies lead to a future where genetic enhancements are the norm? Will there be a divide between those who can afford genetic enhancements and those who cannot? These questions bring to light the need for broader discussions and public engagement on the ethical, social, and philosophical dimensions of gene editing technologies. The accessibility and affordability of gene editing technologies, such as CRISPR, have the potential to revolutionize healthcare and biotechnology, with the potential to cure diseases, change organisms, and reshape the future of humanity. While the increased accessibility and affordability of these technologies can lead to significant advancements and democratize access to healthcare, they also raise important ethical concerns that must be addressed. It is crucial to ensure that these technologies are used responsibly, with rigorous regulatory frameworks in place to protect human rights and dignity. The implications of gene editing technologies extend beyond healthcare, raising questions about their impact on ecosystems, food security, and the very nature of being human. As we venture into the exciting world of genetics and CRISPR, it is essential that we approach these new frontiers with a balanced perspective, considering both the tremendous opportunities and the potential risks they entail.

IMPLICATIONS OF UNEQUAL ACCESS TO GENE EDITING ADVANCEMENTS

The implications of unequal access to gene editing advancements are manifold and complex. One significant consequence is the exacerbation of existing social and economic inequalities. Gene editing technologies have the potential to cure diseases and enhance human capabilities, but if only a select few have access to these advancements, it will only widen the gap between the haves and the have-nots. In a world where genetic enhancements can be used to improve intelligence, physical strength, or attractiveness, individuals without access to these technologies would be at a significant disadvantage, perpetuating social inequalities based on genetic factors. This could have serious implications for social mobility and worsen existing disparities.

Unequal access to gene editing could also result in the rise of a genetic elite, a class of individuals who have undergone extensive genetic enhancements. This could lead to the creation of a privileged group that possesses superior abilities and traits, effectively widening the gap between them and the rest of society. Such a scenario raises ethical concerns regarding fairness, justice, and equality. One could argue that unequal access to gene editing technology undermines the principles of meritocracy and equal opportunity, as success and achievement become increasingly determined by genetic factors rather than individual effort or talent. This could have profound implications for social cohesion and the overall functioning of society.

Unequal access to gene editing could also give rise to new forms of discrimination and stigmatization. As genetic enhancements become more common, those without access to these technologies may face marginalization and social exclusion. They could be seen as genetically inferior or disadvantaged, leading to discrimination and prejudice. This could perpetuate existing forms of discrimination based on race, gender, or socioeconomic status, and contribute to the development of a new form of discrimination known as "genetic discrimination." This could have significant implications for individual well-being and psychological health, as those without access to gene editing technologies may perceive themselves as being inherently inferior or inadequate.

Another profound implication of unequal access to gene editing advancements is the potential for a two-tiered healthcare system. If gene editing technologies become commercialized and available only to those who can afford them, it could result in a healthcare divide where the rich have access to life-enhancing treatments, while the poor are left behind. This could further exacerbate health disparities, as those without access to gene editing technologies may be denied the benefits of genetic enhancements and personalized medicine. This raises important questions about the ethical and moral responsibilities of governments and healthcare systems to ensure equitable access to these advancements. Unequal access to gene editing advancements could also have implications for environmental sustainability and biodiversity. As organisms are genetically modified and altered, there is a risk of unintended consequences and ecological disruptions. If only a select few have access to these technologies, they could manipulate and modify organisms without considering the broader ecological impact. This could lead to the loss of biodiversity, the spread of invasive species, or the creation of genetically modified organisms that are ill-suited to their environment. It is crucial, therefore, to ensure that the benefits and risks of gene editing technologies are balanced and that decision-making processes involving genetic modifications are transparent and inclusive. The implications of unequal access to gene editing advancements are far-reaching and multifaceted. From exacerbating social and economic inequalities, to creating a genetic elite and perpetuating discrimination, to undermining healthcare equality and ecological sustainability, unequal access to gene editing poses significant challenges for society. It is crucial to consider the ethical, social, and environmental implications of these advancements and to ensure that access to gene editing technologies is equitable and just. Only by doing so can we harness the potential of gene editing to benefit all of humanity, rather than a select few.

BALANCING THE ADVANTAGES WITH POTENTIAL DISPARITIES AND DISCRIMINATION

While the advancements in genetics and CRISPR technology offer immense possibilities for the advancement of medicine and the improvement of human lives, it is crucial to address the potential disparities and discrimination that may arise from these break-throughs. One of the primary concerns is the accessibility of these technologies to all segments of society. As gene-editing tech-niques become further developed and refined, it is imperative that they are made accessible to everyone, regardless of their socioeconomic status or geographic location. Given the poten-tially high costs associated with these advanced procedures and treatments, there is a risk that only certain privileged groups might have access to these benefits, while others are left behind. This could exacerbate existing social and health inequalities and lead to further marginalization of certain communities.

Another significant issue to consider is the potential for genetic discrimination. As we gain the ability to modify the human ge-nome, questions arise about the repercussions of such modifica-tions on societal values, norms, and perceptions. Discrimination based on genetic traits could become a reality, where certain in-dividuals are stigmatized or marginalized due to their genetic makeup. This could lead to the creation of a genetic underclass, with individuals who have not benefited from genetic enhance-ments being viewed as inferior or less desirable. This kind of dis-crimination would not only be ethically problematic but could also have detrimental social consequences, further widening divisions within society. There are important ethical considerations when it comes to gene editing and its impact on future generations. While CRISPR technology allows us to make targeted modifications to the DNA of individuals, these changes can also be passed down to future generations through germline editing. This raises questions about the potential long-term effects and unintended consequences of such interventions. We must carefully consider the implications of permanently altering the genetic makeup of our species, as it can have profound and irreversible effects on future generations. It is not only a matter of individual choice but also a collective responsibility to ensure that the choices we make today do not jeopardize the well-being and autonomy of those who come after us.

A comprehensive regulatory framework is necessary to address these potential disparities and discrimination. Governments and international bodies must work together to establish guidelines and standards that ensure the equitable distribution of gene-editing technologies and prevent any form of genetic discrimination. This would require active involvement from policymakers, scientists, and ethicists to strike the delicate balance between scientific progress and social responsibility. A robust system of checks and balances is necessary to prevent abuses of these technologies and safeguard the rights and dignity of all individuals. Education and public awareness also play a crucial role in addressing these concerns. As society becomes more aware of the potential benefits and risks associated with genetics and CRISPR technology, it is important to foster an informed public discourse that allows for a wide range of perspectives to be heard. This would help ensure that decisions about the use of these technologies are made with adequate input and representation from all affected stakeholders. Education programs should aim to promote a nuanced understanding of genetics and gene editing, dispelling misconceptions and myths that could contribute to discrimination and stigmatization.

While the advancements in genetics and CRISPR technology offer tremendous potential for improving human lives, it is crucial to address potential disparities and discrimination that may accompany these breakthroughs. Ensuring equitable access to these technologies, preventing genetic discrimination, and carefully considering the long-term effects on future generations are all vital factors in navigating the exciting and complex world of gene editing. Through a comprehensive regulatory framework and widespread public education, we can strive to reap the benefits of these technologies while safeguarding the fundamental principles of equality and justice. It is our collective responsibility to shape the future of humanity in a way that is inclusive, ethical, and respectful of the diversity of individuals and communities.

The field of genetics has seen tremendous advancements in recent years, with the development of gene editing techniques like CRISPR paving the way for a host of exciting possibilities. CRISPR is a revolutionary gene-editing tool that allows scientists to make precise changes to an organism's DNA. This breakthrough technology has the potential to cure diseases, change the makeup of organisms, and even reshape the future of humanity itself. One of the most promising applications of CRISPR is its potential to cure genetic diseases. Genetic disorders are caused by mutations or abnormalities in an individual's DNA, and they can often lead to severe health problems and reduced quality of life. Traditionally, treating genetic diseases has been a difficult and complex task, with limited success. With the advent of CRISPR, scientists now have a powerful tool at their disposal to directly target and edit the faulty genes responsible for these disorders. Imagine a future where diseases like cystic fibrosis, sickle cell anemia, or Huntington's disease can be cured through a simple genetic intervention. CRISPR has already shown promise in preclinical trials for these and many other genetic disorders. By using this revolutionary technology, scientists have been able to edit the genes of affected cells and correct the underlying genetic causes of these diseases. This offers hope for individuals who currently suffer from genetic disorders and presents a potential cure for future generations. Another fascinating application of CRISPR is its ability to change the DNA of organisms, including plants and animals. This opens up a whole new realm of possibilities, from creating disease-resistant crops to engineering animals with desirable traits. For example, CRISPR could be used to develop genetically modified crops that are more resistant to pests and diseases, reducing the need for harmful pesticides and herbicides. Such advancements in agriculture could help address global food security challenges and contribute to a more sustainable future. The potential to genetically modify animals has far-reaching implications in fields such as medicine and conservation. Scientists can use CRISPR to modify the DNA of animals, making them more resistant to diseases or enhancing their ability to survive in changing environments. This could be particularly beneficial for endangered species, allowing for the preservation and recovery of populations at risk of extinction. Genetically modified animals could be used to produce valuable therapeutic proteins, such as insulin or clotting factors, in a more efficient and sustainable manner. While the possibilities afforded by CRISPR are incredibly exciting, there are also significant ethical and societal considerations that must be addressed. The power to edit genes raises questions about how far we should go in altering the blueprint of life. Issues of consent, equity, and unintended consequences need to be carefully considered and regulated. The ability to modify human embryos, for example, raises complex ethical questions about the morality of changing the genetic makeup of future generations. Striking a balance between scientific progress and responsible innovation will be crucial as we navigate the new frontiers of genetics. The breakthroughs in genetics and the development of CRISPR have opened up a world of possibilities. From curing genetic diseases to modifying the DNA of organisms, these advancements have the potential to reshape the future of humanity. CRISPR offers hope for individuals and families affected by genetic disorders, providing the possibility of a cure and improved quality of life. The ability to change the DNA of organisms presents opportunities for revolutionizing agriculture, conservation, and medicine. The ethical considerations surrounding gene editing must not be overlooked, and careful regulation is necessary to ensure responsible innovation. As we venture to the edge of science, it is imperative that we proceed with caution and Mindfulness, recognizing the immense power and potential of genetics in shaping the future.

XI. UNCHARTED ETHICAL TERRITORIES

As the field of genetics advances at an unprecedented pace, we find ourselves on the precipice of uncharted ethical territories. The advent of gene editing technologies such as CRISPR has raised a myriad of moral dilemmas that demand our immediate attention and consideration. With the power to manipulate the very fabric of life, we must grapple with questions concerning the boundaries of human intervention, the potential repercussions of genetic modifications, and the implications for social inequalities. One of the primary concerns surrounding gene editing and CRISPR technology involves the alteration of germline cells, the genetic material passed on to future generations. While this has the potential to eliminate genetic diseases from the gene pool, it also raises profound ethical questions. By editing the germline, we are fundamentally altering the genetic blueprint of humanity. This raises concerns about the possibility of unintended consequences or unforeseen side effects that may only become apparent in future generations. The pursuit of creating "designer babies" with desirable traits raises concerns regarding eugenics and the potential for creating a society that values certain genetic characteristics above others. Gene editing and CRISPR technology raise questions about consent and autonomy. Who has the authority to make decisions about genetic modifications? Should individuals have the right to alter their own genetic code, or should these decisions be subjected to societal or regulatory oversight? The ability to enhance certain traits or eliminate diseases through gene editing brings forth the possibility of creating a genetic divide between those who can afford such interventions and those who cannot, exacerbating existing social inequalities. Concerns are raised regarding the potential for coercion, as individuals may feel pressured to undergo genetic modifications to fit societal norms or expectations. Another significant ethical consideration lies in the potential misuse of gene editing technology. While its applications for disease treatment and prevention are laudable, there exists the potential for nefarious purposes. For instance, the ability to enhance certain traits for non-medical reasons, such as intelligence or physical appearance, raises concerns about creating a society that places undue emphasis on genetic superiority, leading to discrimination and marginalization. The possibility of biological warfare or the creation of genetically modified organisms with unintended consequences poses a significant risk to global security and ecological balance. The impact of gene editing technology extends beyond human genetics and raises questions about our responsibility towards the natural world. CRISPR allows for the modification of not only human genes but also those of other organisms. This opens up a realm of possibilities, from eliminating pests and invasive species to enhancing the productivity of agricultural crops. The ethical implications of such developments cannot be understated. The potential unintended consequences, including the disruption of ecosystems and loss of biodiversity, warrant cautious consideration and assessment of risks and benefits before carrying out genetic modifications on non-human organisms.

In navigating these uncharted ethical territories, it is paramount that we adopt a multidisciplinary approach that includes scientists, ethicists, policymakers, and the public. Open and transparent discussions must take place to ensure that the potential benefits of gene editing and CRISPR technology are balanced with the ethical considerations they raise. International cooperation and collaboration are crucial in establishing guidelines and regulations that govern the responsible use of these powerful tools. Robust ethical frameworks need to be developed and continually evaluated and updated in response to emerging technologies and societal changes. The advancements in gene editing technology, particularly CRISPR, propel us towards uncharted ethical territories that demand our immediate attention. From the altering of germline cells to the potential for genetic discrimination and the implications for the natural world, the ethical dilemmas we face are complex and multifaceted. As we venture further into this exciting world of genetics, it is imperative that we navigate these uncharted territories with caution, responsibility, and an unwavering commitment to the well-being of humanity and the planet.

ETHICAL CONSIDERATIONS OF GERMLINE GENE EDITING

Germline gene editing, the ability to make permanent changes to the genetic makeup of future generations, presents a myriad of ethical considerations. The potential to eliminate harmful genetic diseases and enhance desired traits is undeniably appealing, but these advancements also raise concerns about the ethical boundaries of altering the human genome. One primary concern is the question of consent. Germline gene editing involves making modifications at the embryonic stage, before an individual is able to provide informed consent for such interventions. This raises significant ethical questions about the right to autonomy and the potential for infringements upon an individual's genetic heritage. Another ethical consideration is the potential for unintended consequences. As our understanding of genetics and gene editing techniques continues to evolve, there remains a considerable risk of unforeseen genetic mutations and unintended side effects. The long-term effects of gene editing within the germline are still relatively unknown, and the possibility of creating unintended genetic abnormalities or altering the natural evolutionary trajectory of the human species raises considerable concerns over the potential for irreparable harm to future generations.

The concept of germline gene editing also raises important questions around equity and social justice. If these technologies become widely available, there is a risk that they may only be accessible to the wealthy and privileged, further exacerbating existing social inequalities. There is the potential for disparities in access to genetic enhancements, creating a divided society based on genetic advantages and disadvantages. This raises concerns about creating a genetic elite and widening the gaps between social classes, potentially leading to a dystopian future where genetic superiority determines one's worth.

The implications for eugenics cannot be ignored. While the aim of germline gene editing is to improve the human condition and eradicate genetic diseases, the potential for misuse and abuse of this technology must be carefully considered. The historical associations of eugenics with ideologies of racial superiority and forced sterilization raise concerns about the potential for germline gene editing to be used to perpetuate discrimination and inequality. There is a critical need for a comprehensive ethical framework to ensure that the use of this technology is guided by principles of equality, diversity, and respect for human rights. The ethical implications of germline gene editing also extend beyond the realm of humans. With the ability to edit the genetic code of organisms, there is the potential for widespread ecological impacts. Altering the genetic makeup of species could inadvertently disrupt ecosystems, leading to unintended consequences for biodiversity and the stability of natural environments. The responsibility for ensuring the ethical use of this technology extends to considerations of the broader ecological impact and the potential for irreversible harm to the natural world. In addition to these concerns, there are also questions about the slippery slope argument. Some argue that germline gene editing may lead to a future where parents have the ability to select for specific traits in their children. This raises fears of a society driven by the pursuit of perfection, where diversity and individuality are diminished in favor of conformity to predetermined genetic standards. The potential loss of genetic variation and the impacts on the overall resilience and adaptability of the human species are important considerations when evaluating the ethical implications of germline gene editing.

Germline gene editing holds immense promise for the future of humanity, but it also presents complex ethical considerations that demand careful discourse and regulation. The potential to eradicate genetic diseases and enhance desirable traits must be balanced with concerns around consent, unintended consequences, equity, eugenics, ecological impacts, and the erosion of diversity. As we venture into this new frontier of genetics, it is vital that we establish a robust ethical framework to guide the responsible and equitable use of these technologies. Only then can we harness the potential of germline gene editing to shape a future that is both scientifically advanced and ethically just.

THE ROLE OF INTERNATIONAL COLLABORATION AND REGULATING GENE EDITING

The role of international collaboration and regulating gene editing is crucial in order to ensure responsible and ethical use of this powerful technology. Gene editing has the potential to revolutionize medicine and agriculture, but it also raises serious ethical concerns. International collaboration is necessary to establish guidelines and protocols that can govern the use of gene editing techniques such as CRISPR. By working together, scientists and policymakers from different countries can share knowledge, expertise, and best practices, and ensure that the field of gene editing progresses in a responsible manner. International collaboration can help ensure that the benefits and risks associated with gene editing are fairly distributed among different countries and communities, and prevent a deepening of existing inequities. For example, it is important to ensure that gene editing is accessible and affordable to people in developing countries, and that their voices are included in discussions about the ethics and governance of gene editing. Regulating gene editing is also critical to protect against potential misuse or abuse of this technology. Enforcing strict regulations and controls can help prevent the creation of dangerous genetically modified organisms, and ensure that gene editing is used only for beneficial purposes. A global regulatory framework can help establish standards for safety, efficacy, and ethical considerations in gene editing research and applications. It can also help facilitate the exchange of information and resources among countries, and promote transparency and accountability in the field of gene editing. Regulation can help address concerns about genetic discrimination and eugenics, by ensuring that gene editing is not used to create "designer babies" or to unfairly manipulate human traits.

It is important to strike a balance between regulation and innovation in the field of gene editing. Excessive regulation could stifle scientific progress and hamper the potential of gene editing to address pressing global challenges such as disease, hunger, and climate change. It is therefore crucial to have a regulatory framework that supports responsible innovation and allows for continued scientific discovery in gene editing. This can be achieved through a combination of government regulation, self-regulation within the scientific community, and public engagement and input. The involvement of diverse stakeholders such as scientists, bioethicists, policymakers, patient advocacy groups, and the general public is essential to ensure that decisions about the regulation of gene editing are informed by a wide range of perspectives and take into account the broader societal implications. In addition to international collaboration and regulation, public education and dialogue are key to fostering an informed and ethical approach to gene editing. As this technology becomes more accessible and its applications more widespread, it is important that the general public understands the science behind gene editing, as well as its potential benefits and risks. Public engagement can help address concerns, dispel myths, and build public trust in the responsible use of gene editing. It can also help broaden the conversation about the ethical, social, and legal implications of gene editing beyond the scientific community, and involve the public in decision-making processes related to gene editing regulation. International collaboration and regulation play a critical role in the responsible and ethical use of gene editing. By working together, countries can establish guidelines and protocols that govern the use of gene editing techniques such as CRISPR. Regulation is necessary to protect against potential misuse or abuse of gene editing and ensure that the benefits and risks associated with this technology are fairly distributed. It is important to strike a balance between regulation and innovation, and to involve diverse stakeholders in decision-making processes. Public education and dialogue are also crucial to foster an informed and ethical approach to gene editing and to build public trust in its responsible use. As we venture into the exciting world of genetics and CRISPR, it is imperative that we navigate these new frontiers with careful consideration and a commitment to the well-being of humanity.

RESPONSIBLE AND TRANSPARENT APPROACHES TO GENE-EDITING RESEARCH

They are essential in order to navigate the ethical and social implications of this groundbreaking technology. As gene-editing techniques, such as CRISPR, become more refined and accessible, it is imperative that scientists adhere to strict ethical guidelines and engage in open dialogue with the public. One of the key considerations in responsible gene-editing research is the potential for unintended consequences or off-target effects. By implementing rigorous testing and quality control measures, researchers can mitigate the risks associated with gene editing and ensure that any modifications made are intentional and precise. Transparency in the research process is critical to building public trust and ensuring that decisions regarding gene editing are made collectively and with the input of diverse stakeholders.

The responsible use of gene-editing techniques also entails careful consideration of the impact on human health and safety. Before any gene-editing technologies are implemented in clinical settings, they must undergo rigorous preclinical and clinical testing. This process not only establishes the safety and efficacy of the technique but also allows for thorough evaluation of its potential long-term effects. Clinical trials must be conducted in a transparent manner, with clear protocols and guidelines, in order to ensure that the risks and benefits of gene editing are fully understood by both researchers and participants. Ongoing monitoring and evaluation of patients who have undergone gene editing will be crucial in order to assess any unforeseen consequences or long-term effects. Transparency and responsible accountability also extend to the commercialization and accessibility of gene-editing technologies. As these techniques have the potential to revolutionize healthcare and disease prevention, it is important to ensure that they are not only available to those who can afford them. Steps should be taken to prevent the patenting of basic gene-editing technologies, which would limit their accessibility and hinder research progress. Instead, partnerships between academia, industry, and government should be fostered to ensure that gene-editing technologies are affordable, widely available, and used for the benefit of all. In addition to practical considerations, responsible gene-editing research necessitates careful ethical and social reflection. Gene editing raises profound questions about our understanding of human nature, the boundaries of intervention, and the potential for unintended consequences. These questions must be addressed in an inclusive manner, involving input from diverse perspectives, including scientists, ethicists, policymakers, and the public. Public engagement and open dialogue are key in order to ensure that the decisions regarding gene editing are made collectively and reflect our shared values and aspirations. Responsible gene-editing research requires a commitment to equity and justice. This technology has the potential to exacerbate existing societal disparities if it is not implemented in a fair and equitable manner. Special attention must be paid to issues of genetic discrimination, access to healthcare, and the potential for eugenics-like practices. Safeguards should be put in place to prevent the misuse of gene editing for unethical purposes and to prevent discrimination based on genetic information. Regulations and oversight mechanisms should be established to ensure that gene-editing technologies are used responsibly and for the public good.

Responsible and transparent approaches are imperative in gene-editing research to navigate the ethical, social, and practical challenges that arise with this groundbreaking technology. By adhering to strict ethical guidelines, conducting rigorous testing and monitoring, promoting transparency, fostering accessibility, encouraging public engagement, and upholding principles of equity and justice, we can harness the power of gene editing to improve human health while minimizing the risks and ensuring that decisions regarding gene editing are made collectively and reflect our shared values. Only through responsible and transparent practices can we fully realize the potential of gene editing and ensure that it contributes to a brighter and more equitable future for humanity. The field of genetics has seen tremendous advancements in recent years, with the emergence of a revolutionary technology known as CRISPR. This gene-editing technique has unlocked endless possibilities in the realm of genetic manipulation, offering scientists the ability to alter an organism's DNA with incredible precision. CRISPR holds the potential to cure diseases, modify crops for increased yield and resistance to pests, and even reshape the future of the entire human race. The implications of this groundbreaking technology are profound and far-reaching, encompassing both the promise of tremendous benefits and a host of ethical and moral concerns. As we embark on this journey to the edge of science, it becomes clear that the new frontiers of genetics will shape the world in ways we never thought possible. One of the most remarkable aspects of CRISPR is its ability to cure diseases that were once thought to be incurable. With this gene-editing tool, scientists can target and correct specific mutations in an individual's DNA, effectively eliminating the root cause of various genetic disorders. Diseases such as cystic fibrosis, sickle cell anemia, and even certain types of cancer could potentially be eradicated through the use of CRISPR. The possibilities for personalized medicine are unprecedented, as treatments can be tailored to an individual's unique genetic makeup. The dream of a world free from the burden of genetic diseases is inching closer to reality, thanks to the groundbreaking advancements in genetics and CRISPR.

CRISPR has the potential to transform the agricultural industry by creating genetically modified crops that are more resilient and productive. By introducing specific genetic modifications into plants, scientists can enhance their ability to withstand drought, combat pests and diseases, and increase their nutritional value. This technology offers a potential solution to global food security challenges, as crops can be engineered to thrive in harsh environments and yield higher quantities of nutritious food. The implications for developing countries, where hunger and malnutrition are rampant, are both profound and life-saving. CRISPR holds the power to address pressing global issues and reshape the way we produce food, ensuring a sustainable future for generations to come. As with any groundbreaking technology, CRISPR comes with a host of ethical and moral considerations. The ability to manipulate an organism's DNA raises questions about the boundaries of science and the potential for unintended consequences. The capacity to alter the genetic blueprint of future generations has raised concerns about the concept of designer babies and the potential for creating a genetically superior elite. There is a fine line between curing genetic diseases and crossing into the realm of eugenics, which poses a threat to our understanding of equality and the value of human life. As we delve deeper into the new frontiers of genetics, it is crucial to engage in ethical and moral discussions to ensure responsible use of this powerful tool. In addition to the ethical dilemmas surrounding CRISPR, there are also concerns about the long-term effects of genetic manipulation on biodiversity and the environment. Altering the genetic makeup of organisms can have unintended consequences on ecosystems, as we cannot fully predict the interconnectedness of living organisms. The introduction of genetically modified crops, for example, may disrupt natural pollination patterns or create superweeds resistant to herbicides. It is imperative to carefully consider the potential risks and implement robust regulatory frameworks to mitigate any adverse effects on the environment and biodiversity.

The new frontiers of genetics, propelled by the breakthrough technology of CRISPR, offer immense promise for curing diseases, transforming agriculture, and shaping the future of humanity. The ability to edit an organism's DNA with precision has opened up a world of possibilities that were once the stuff of science fiction. As we venture into this exciting realm of science, it is essential to approach it with caution and ethical responsibility. The potential for medical breakthroughs and solving global challenges is immense, but so are the potential risks and ethical dilemmas. The journey to the edge of science requires us to navigate these uncharted territories with wisdom and foresight, ensuring that the benefits of genetic manipulation are realized without compromising our values and the sanctity of life.

XII. UNFORESEEN CONSEQUENCES

While the potential of gene editing and CRISPR to revolutionize medicine and reshape the future of humanity is undoubtedly exciting, we must also acknowledge the potential for unforeseen consequences that may arise from these breakthrough technologies. One of the main concerns surrounding gene editing is the ethical implications it presents. As we gain the ability to modify the genetic code of organisms, we are essentially playing the role of "creator," which raises questions about the boundaries of human interference in the natural order of things. The idea of germline editing, where changes made to an individual's DNA can be passed on to future generations, presents complex ethical dilemmas. The decisions we make today about editing the human genome could have far-reaching effects on future generations, leading to unintended and irreversible consequences.

Another unforeseen consequence of gene editing is the potential for unintended genetic mutations. While CRISPR-Cas9 is highly precise, there is still the possibility of off-target effects, where genetic modifications occur in unintended locations in the genome. These off-target effects could lead to harmful mutations, potentially causing new diseases or exacerbating existing ones. The long-term effects of gene editing are not yet fully understood, as changes to the genetic code can have cascading effects on an organism's overall health and development. This lack of complete understanding highlights the need for continued research and cautious implementation of these technologies to minimize any potential negative impacts.

In addition to ethical and health concerns, gene editing also has the potential to exacerbate existing social inequalities. As with many new technologies, the initial cost of gene editing procedures may be prohibitively high, limiting access to those who can afford it. This could create a divide between those who have the means to enhance their genetic makeup and those who do not, leading to an increased disparity in the capabilities and opportunities available to different segments of society. The potential for "designer babies," where parents could select specific traits for their children, may further reinforce societal hierarchies based on perceived genetic superiority, potentially eroding concepts of diversity and equality. The use of gene editing in agriculture and the environment may have unintended consequences for ecosystems and biodiversity. While the promise of genetically modified crops and animals may lead to increased food production and resistance to diseases or environmental conditions, it could also disrupt ecosystems by introducing invasive species or creating imbalances in natural predator-prey relationships. The long-term effects on biodiversity are uncertain, as genetically modified organisms could outcompete or replace native species, leading to a loss of ecological diversity and resilience.

There is the concern that the rapid development of gene editing technologies may outpace the ability of society to effectively regulate their use. As breakthroughs continue to occur at an astonishing pace, frameworks and guidelines for responsible implementation and oversight may lag behind. Without comprehensive regulation, gene editing could be used in irresponsible or unsafe ways, potentially leading to unforeseen consequences that could harm individuals or society as a whole. It is crucial that policymakers, scientists, and ethicists work together to establish clear guidelines and ethical boundaries to ensure the responsible and equitable use of gene editing technologies.

The exciting world of genetics and CRISPR presents incredible possibilities for the future of humanity. We must approach these breakthroughs with caution and consider the potential for unforeseen consequences. The ethical implications, risks of unintended mutations, exacerbation of social inequalities, threats to ecosystems and biodiversity, and challenges of regulation all demand careful consideration. By acknowledging these complexities and taking a thoughtful and cautious approach, we can navigate the new frontiers of genetics in a way that maximizes benefits and minimizes harm, paving the way for a future that is both scientifically advanced and ethically sound.

POTENTIAL RISKS OF GENETIC MODIFICATIONS

While the advances in genetics and gene editing techniques like CRISPR hold tremendous promise for humanity, they also come with a host of potential risks that cannot be overlooked. One major concern is the possibility of unintended consequences resulting from genetic modifications. When scientists manipulate an organism's DNA, they often do so with a specific goal in mind, such as curing a disease or enhancing certain traits. The intricate nature of genetics means that there can be unforeseen effects that manifest in unexpected ways. For example, altering a gene related to disease resistance in a plant may inadvertently affect its ability to reproduce or interact with other organisms in its ecosystem. These unintended consequences could have far-reaching implications, not only for the modified organism but also for the entire ecosystem it belongs to. Another risk associated with genetic modifications is the potential for genetic discrimination. As genetic testing becomes more widespread and accessible, there is a concern that individuals could be unfairly treated based on their genetic makeup. Employers, insurance companies, or even potential romantic partners may use genetic information to make decisions about hiring, coverage, or relationships, leading to discrimination and inequality. This issue is particularly pertinent when considering genetic modifications in humans, where the ability to select or alter traits could give rise to a society fragmented along genetic lines. There is the potential for misuse of genetic information by governments or other authoritative bodies, leading to violations of privacy and individual autonomy.

Ethical concerns also surround the idea of genetic modifications. The ability to edit genes has raised questions about the moral boundaries of science and how far researchers should go in their pursuit of genetic advancements. Traditional moral frameworks may need to be reevaluated as the implications of these technologies become more apparent. For example, should we allow genetic modifications that enhance physical or cognitive abilities, potentially exacerbating existing social inequalities? Should we edit embryos to eradicate fatal diseases, potentially opening the door to "designer babies"? These are complex ethical questions that require careful consideration and public discourse.

The scope for unintended consequences and ethical dilemmas is amplified by the global spread of genetic modifications. Due to the interconnectedness of the world, the actions of one country or institution in genetic research can have implications for others. The release of genetically modified organisms into the environment without adequate testing or understanding of potential risks could lead to ecological disasters or unforeseen impacts on local communities. If genetic modifications become widely available, disparities in access and affordability could exacerbate existing socioeconomic inequalities on a global scale, creating a divide between those who can afford genetic enhancements and those who cannot. The potential for bioterrorism is another significant risk associated with genetic modifications. As gene editing techniques become more accessible, the concern arises that these technologies could be misused by individuals or groups seeking to cause harm. The ability to engineer and release genetically modified pathogens or create untreatable diseases raises serious national security concerns. The misuse of genetic modifications could potentially lead to devastating

consequences that are difficult to contain or combat, further ne-
cessitating the need for strict regulation and oversight.

While genetic modifications and the advancements in gene editing techniques like CRISPR hold immense promise for the future
of humanity, they also come with a range of potential risks that
must be addressed. The unpredictable nature of genetics can result in unintended consequences, while the widespread adoption
of genetic modifications raises concerns about genetic discrimination and the erosion of privacy. Ethical considerations and
global implications further underscore the need for thoughtful
regulation and public discourse surrounding these technologies.
The potential for bioterrorism poses a grave threat that cannot
be overlooked. As we journey further into the exciting world of
genetics, it is crucial that we navigate these risks responsibly and
with a deep understanding of the potential consequences.

ECOLOGICAL DISRUPTIONS AND UNINTENDED EFFECTS ON ECOSYSTEMS

Ecological disruptions and unintended effects on ecosystems are some of the concerns that arise with the advancements in genetics and CRISPR technology. While gene editing holds the promise of curing diseases and transforming organisms, it is essential to examine its potential impact on the environment. The modification or removal of certain genes within an organism might inadvertently disrupt the delicate ecological balance within a given ecosystem. The introduction of genetically modified organisms (GMOs) into the environment may result in unintended consequences that can have far-reaching effects on ecosystems. These disruptions can occur through various mechanisms, including the alteration of food chains, the spread of modified genes to non-target species, and the potential for gene flow between GMOs and related wild populations. One area of concern is the alteration of food chains due to genetic modifications. Introducing genetic modifications into an organism can impact its feeding habits, which can subsequently affect the organisms that depend on it for food. For example, if a genetically modified plant species becomes resistant to pests, it may result in a decline in the population of a particular insect that feeds on that plant. This reduction in insect population can have cascading effects on other species that rely on that insect for sustenance, potentially disrupting the overall ecological balance of the ecosystem.

Another potential disruption arises from the spread of modified genes to non-target species. GMOs can reproduce and pass on their modified genes to wild populations, potentially altering the genetic composition and characteristics of these populations. This genetic flow can occur through unintentional crossbreeding between GMOs and wild relatives, allowing modified genes to enter the wild gene pool. This situation raises concerns regarding the long-term consequences of such genetic mixing, as it could result in the unintended alteration of traits in non-target species. The spread of modified genes can lead to the loss of genetic diversity within wild populations, reducing their ability to adapt to changing environmental conditions. The potential for gene flow between GMOs and related wild populations can cause unintended effects on ecosystems. Gene flow occurs when genes from one population mix with genes from another population through interbreeding. In the case of GMOs, gene flow can lead to the transfer of modified genes from genetically engineered organisms to wild populations. This transfer of genes can have unpredictable consequences, as it can result in the spread of genetically modified traits and characteristics in the natural environment. This situation raises questions about the potential for genetic pollution and the loss of natural biodiversity as modified traits become more prevalent in wild populations. The ability of GMOs to outcompete their wild counterparts due to their modified traits can lead to the displacement or even extinction of native species, further exacerbating the potential ecological disruptions caused by gene flow. To mitigate the ecological disruptions and unintended effects on ecosystems, careful consideration and regulation of genetic modifications are essential. Rigorous testing and risk assessment procedures should be in place to evaluate the potential ecological impacts of any genetically modified organism before its release into the environment. These evaluations should include a thorough analysis of the organism's potential interactions with its environment, including its role within food chains and the likelihood of gene flow to non-target species. Measures should be taken to minimize the likelihood and consequences of unintended gene flow, such as physical barriers or containment strategies. Responsible and transparent decision-making processes should involve scientific experts, policymakers, and representatives from relevant stakeholders to ensure the careful consideration of the potential ecological effects of genetic modifications. While gene editing and CRISPR technology offer exciting opportunities for curing diseases and transforming organisms, it is crucial to consider the potential ecological disruptions and unintended effects on ecosystems. Alterations to food chains, the spread of modified genes to non-target species, and gene flow between GMOs and wild populations can all have far-reaching consequences on the delicate balance of ecosystems. By implementing robust risk assessment procedures and responsible decision-making processes, we can strive to ensure the safe and ethical use of genetic modifications, minimizing the potential ecological disruptions and safeguarding our ecosystems for future generations.

THE IMPORTANCE OF THOROUGH RESEARCH AND RISK ASSESSMENT BEFORE IMPLEMENTING GENE EDITING TECHNIQUES

In the ever-evolving field of genetics and the emergence of gene editing techniques, it is imperative to conduct thorough research and risk assessment before implementing these groundbreaking technologies. Gene editing holds immense potential to cure diseases, alter organisms, and potentially reshape the future of humanity. The significance of careful research and risk assessment cannot be overstated in ensuring the safe and ethical utilization of gene editing techniques. Firstly, thorough research is critical in understanding the complexities of gene editing and its potential ramifications. As gene editing techniques such as CRISPR-Cas9 gain traction in the scientific community, the need for a comprehensive understanding of their mechanisms becomes crucial. Researchers must thoroughly investigate the various intricacies involved in the gene editing process, including target specificity, off-target effects, and unintended consequences. By delving deep into the underlying principles of gene editing, scientists can anticipate potential risks and develop strategies to mitigate them. Thorough research contributes to the accumulation of knowledge and expertise, enabling scientists to refine gene editing techniques over time and maximize their efficiency and safety. Without a strong foundation of scientific understanding, the implementation of gene editing techniques without proper research could lead to dire consequences, both for individuals undergoing gene editing procedures and for society as a whole.

A meticulous risk assessment is essential to ensure the ethical and responsible use of gene editing techniques. Gene editing has the power to transform the lives of individuals suffering from genetic disorders and to address various societal challenges. The potential risks associated with gene editing cannot be overlooked. It is paramount to conduct a thorough assessment of the potential ethical dilemmas, social implications, and unintended consequences that arise from implementing gene editing techniques. By carefully weighing the risks and benefits, scientists, policymakers, and society at large can make informed decisions regarding the application of these technologies. This risk assessment should involve cross-disciplinary collaboration, including input from experts in bioethics, law, sociology, and other relevant fields. Engaging the public in the conversation surrounding gene editing is essential to ensure democratic decision-making and to address any ethical concerns or reservations. Responsible assessment of risks allows for an implementation process that is mindful of potential harms and strives to maximize benefits.

Conducting extensive research and thorough risk assessment in the field of gene editing helps to avoid hasty, ill-informed decisions with far-reaching consequences. In the face of the excitement and promise surrounding gene editing technologies, it is crucial to resist the temptation of rushing into implementation without proper evaluation. Without comprehensive research and risk assessment, the potential dangers and unanticipated consequences of gene editing may remain unrecognized until it is too late. A cautious and measured approach is vital in order to identify and address any potential pitfalls before they become widespread. This ensures that gene editing techniques are used responsibly and ethically, avoiding any unnecessary harm that may arise from impulsive actions. The importance of thorough research and risk assessment before implementing gene editing techniques cannot be emphasized enough. The potential of gene editing to cure diseases, alter organisms, and reshape humanity is undoubtedly exciting. To maximize the benefits and ensure the safe and ethical use of gene editing, a robust research foundation and meticulous risk assessment are indispensable. By deeply understanding the complexities of gene editing and the potential risks involved, researchers can develop strategies to mitigate these risks. Conducting comprehensive risk assessments enables society to make informed and responsible decisions when it comes to gene editing implementation. By taking these precautionary measures, we can harness the transformative power of gene editing while minimizing the potential harms and preserving the ethical fabric of society. Gene editing and CRISPR technology have emerged as powerful tools in the field of genetics, heralding a new frontier in scientific discovery. These breakthroughs hold the potential to cure diseases, alter organisms, and reshape the future of humanity. The discovery of CRISPR has revolutionized the field of genetics with its unparalleled precision and efficiency. Unlike previous gene-editing techniques, CRISPR allows scientists to target specific genes within an organism's DNA and modify or insert desired genetic material. This newfound ability to manipulate genes with such accuracy has opened up endless possibilities for medical advancements and beneficial genetic modifications. One of the most promising applications of gene editing and CRISPR is in the field of medicine. With CRISPR, scientists can potentially cure genetic diseases that were once thought to be incurable. By specifically targeting faulty genes responsible for these diseases, researchers can edit them to restore their normal function. This offers hope to countless individuals suffering from genetic disorders such as cystic fibrosis, sickle cell anemia, and muscular dystrophy. CRISPR has the potential to eradicate certain infectious diseases by modifying or disabling the genes that enable their transmission. Diseases like HIV, malaria, and Zika virus may no longer pose a threat to global health with the advancements in gene editing technology.

Beyond medicine, gene editing and CRISPR technology have the potential to reshape entire ecosystems and agriculture. By modifying the genes of plants and animals, scientists can enhance their resilience to environmental stressors, increase crop yields, and develop more nutritious food sources. This could have profound implications for food security in a rapidly growing global population. Genetic modifications could also help combat climate change by engineering plants to capture and store more carbon dioxide, mitigating greenhouse gas emissions. Gene editing could aid in the conservation of endangered species by preserving their genetic diversity and enhancing their ability to adapt to changing environments. While the possibilities of gene editing and CRISPR technology are undoubtedly exciting, concerns surrounding ethics and unintended consequences have emerged. The ability to alter the genetic makeup of living organisms raises ethical questions about the boundaries of science and our responsibility as stewards of life. Are we playing God by tampering with the very fabric of life itself? Should we be allowed to modify the genes of future generations, potentially altering the course of human evolution? These questions call for careful consideration and regulation of gene editing practices.

The unintended consequences of genetic modifications cannot be overlooked. Although CRISPR is highly accurate, there is always a risk of unintended off-target effects. Modifying certain genes may have unforeseen consequences on the organism as a whole or its interactions with the environment. The long-term effects of genetic modifications on future generations remain largely unknown. Genetic alterations made today may have far-reaching consequences for future generations, both positive and negative. Careful and rigorous testing should be conducted to assess the safety and potential consequences of gene editing before widespread adoption. Gene editing and CRISPR technology represent a new frontier in scientific discovery that holds immense potential for curing diseases, reshaping ecosystems, and revolutionizing agriculture. The precision and efficiency of CRISPR have opened up a world of possibilities that were once unimaginable. The ethical considerations and unintended consequences associated with these breakthroughs cannot be ignored. It is essential that we approach gene editing with caution, carefully weighing the potential benefits against the possible risks. The future of genetics is undoubtedly exciting, but it requires responsible and ethical scientific practices to ensure that we navigate these new frontiers wisely and responsibly.

XIII. PUBLIC PERCEPTION AND ACCEPTANCE

The public perception and acceptance of gene editing and CRISPR technology play a crucial role in shaping the future of these groundbreaking advancements in genetics. While scientists and researchers continue to push the boundaries of scientific progress in this field, public opinion can significantly influence the direction these technologies take. At present, public perception of gene editing and CRISPR remains mixed. On one hand, there is a sense of excitement and wonder about the potential to cure previously incurable diseases and eliminate genetic disorders from the population. The idea of altering the very building blocks of life to eradicate suffering is undoubtedly appealing. The prospect of using CRISPR technology to enhance certain desirable traits or create genetically modified organisms for agricultural purposes offers a glimpse into a future of limitless possibilities.

On the other hand, public perception is also tempered by concerns and ethical dilemmas. The idea of manipulating human genes raises questions about the limits of scientific intervention and our understanding of the complexities of the human genome. There are concerns about unintended consequences, both for individuals and for society as a whole. The potential for misuse or abuse of these technologies is also a cause for concern. As with any powerful tool, the responsible and ethical use of gene editing and CRISPR is of paramount importance.

The highly technical nature of gene editing and CRISPR technology can create a disconnect between the scientific community and the general public. Difficult concepts and jargon can be hard to grasp for those without a strong background in genetics. As a result, there is a need for effective communication and public engagement to bridge this gap and foster a more informed and inclusive dialogue. Public perception and acceptance can also be influenced by cultural, religious, and societal factors. Different populations may have different views on the moral implications of manipulating genes, based on their values and beliefs. Some cultures may view gene editing as playing God, while others may be more open to the possibilities it presents. Religion, in particular, has a significant influence on public perception. The Catholic Church, for example, has expressed concerns about the potential for gene editing to cross ethical boundaries, while other religious groups may have different perspectives.

Education and awareness are essential in shaping public perception and acceptance. A well-informed public is more likely to understand the potential benefits, risks, and ethical considerations associated with gene editing and CRISPR. By providing accessible and accurate information, governments, scientific institutions, and other stakeholders can help to promote informed decision-making and stimulate discussions about the future direction of these technologies. Public engagement should also include opportunities for dialogue, discussion, and debate. By involving the wider public, including those from diverse backgrounds and perspectives, in the conversation, a more comprehensive understanding of the issues at hand can be developed. Ethical considerations, social impacts, and potential risks need to be carefully weighed against potential benefits, and the public should have a say in determining the boundaries of acceptable use. Public perception and acceptance will shape the future of gene editing and CRISPR. It is important to strike a balance between caution and progress, ensuring that responsible and sustainable practices are pursued while allowing for the potential benefits of these technologies to be realized. By fostering an informed and inclusive dialogue, society can navigate the ethical and moral implications of gene editing, while maximizing the potential benefits for individuals and humanity as a whole. In this way, public perception and acceptance become crucial components in shaping the new frontiers of genetics and the future of humanity.

PUBLIC OPINION ON GENE EDITING

Public opinion on gene editing is a highly complex and contested issue. Gene editing, particularly through the CRISPR technology, has the potential to revolutionize the field of medicine and alter the course of human evolution. The moral and ethical implications of such advancements have sparked a robust debate among scientists, policymakers, and the general public. On one hand, proponents argue that gene editing holds the promise of eradicating genetic diseases and improving overall human health. They point to success stories in the field, such as curing rare genetic disorders and developing more resilient crops. These advancements, they argue, have the potential to alleviate human suffering and create a healthier and more genetically diverse population. Gene editing could potentially allow individuals to optimize certain traits, such as intelligence or athleticism, raising questions about equality and fairness. On the other hand, critics raise concerns about the potential misuse and unintended consequences of gene editing. They argue that altering the genetic makeup of organisms, including humans, raises profound ethical and moral questions about the boundaries of science and our role in shaping the natural world. There are also worries that gene editing could exacerbate existing social inequalities, as those with access to the technology could enhance their genetic traits, creating a new form of genetic elitism. There are concerns over the unforeseen environmental impact of releasing genetically modified organisms into the wild, with potential consequences for ecosystems and biodiversity. The public's opinion on gene editing is shaped by these differing perspectives and is likely influenced by cultural, religious, and personal beliefs. Surveys have found that the general public tends to support gene editing for medical purposes, particularly in the context of treating serious diseases or disabilities. Support decreases when it comes to enhancements or modifying non-disease-related traits. This suggests a nuanced view among the public, weighing the potential benefits against ethical concerns and the possible consequences of a world where gene editing becomes routine. It is clear that public opinion on gene editing is far from monolithic, with differences observed across demographic groups and varying levels of scientific literacy. For instance, individuals with higher levels of education tend to be more accepting of gene editing, likely due to a greater understanding of the science behind it. Conversely, religious beliefs can also play a significant role in shaping attitudes towards gene editing. Some religious groups see gene editing as an affront to nature or a challenge to divine creation, leading to a more skeptical stance. The media's portrayal of gene editing can influence public opinion, as headlines highlighting sensationalistic possibilities or potential dystopian outcomes can shape perceptions of the technology. Understanding public opinion on gene editing is crucial for policymakers and scientists who aim to develop regulations and guidelines that reflect societal values. As gene editing technologies continue to advance and become more accessible, it is essential to engage the public in discussions about the moral, ethical, and social implications of these advancements. Public education and dialogue can help bridge gaps in understanding, foster transparency, and ensure that gene editing is used responsibly and for the collective benefit. Shaping public opinion on gene editing requires an informed and inclusive approach that considers the perspectives of diverse stakeholders. By considering the potential risks and benefits of gene editing, as well as the values and concerns of the public, we can navigate the future of this groundbreaking technology in a way that aligns with our shared vision of a healthier, more just, and sustainable future.

ADDRESSING MISINFORMATION AND INCREASING PUBLIC AWARENESS

While the potential of gene editing and CRISPR technology is indeed thrilling, it is accompanied by a pressing need to address misinformation and increase public awareness regarding these advancements. With the rapid pace at which information spreads in today's digital age, it becomes crucial to ensure that the public has access to accurate and reliable knowledge about the capabilities and limitations of gene editing. Misinformation can lead to fear, misunderstanding, and misguided public opinion, which can hinder progress in the field and impede the responsible and ethical implementation of gene editing.

One key aspect of addressing misinformation is through effective communication and transparency. Experts in genetics and CRISPR technology must engage in clear and accessible dialogue with the public to not only disseminate accurate information but also to address concerns, dispel myths, and emphasize the importance of responsible research and application. This can be achieved through various means like public lectures, educational initiatives, and collaborations with media outlets to ensure accurate representation of scientific advancements. Engaging with policymakers and regulators can help establish guidelines and regulations that balance scientific progress with ethical considerations, ensuring a responsible and well-informed approach to gene editing. Increasing public awareness about genetics and CRISPR technology opens the door for constructive and inclusive conversations about the ethical implications and societal impact of these advancements. It is crucial to involve a wide range of stakeholders, including the general public, patient advocacy groups, bioethicists, and religious leaders, in these discussions to foster a comprehensive understanding of the opportunities and challenges presented by gene editing. By embracing diverse perspectives, we can ensure that decisions regarding the usage of these technologies are made with consideration for the multifaceted needs and concerns of society. In addition to addressing misinformation and fostering inclusive conversations, there is a need to prioritize education and scientific literacy on the subject of genetics and CRISPR technology. Integrating these topics into educational curricula at various levels, from primary schools to undergraduate and postgraduate studies, can help equip future generations with the knowledge and critical thinking skills required to navigate the complexities of gene editing. By investing in science education and encouraging curiosity, we can nurture a scientifically literate society that can actively engage with and contribute to the ongoing dialogue surrounding genetics and CRISPR technology. The dissemination of reliable and understandable information should extend beyond traditional educational settings. Science museums, public exhibitions, and outreach programs can serve as valuable platforms for enhancing public understanding of genetics and CRISPR technology. These interactive and engaging spaces offer opportunities for individuals to explore, ask questions, and interact with experts, thus bridging the gap between scientific research and public engagement. By making science more accessible and relatable, we can empower individuals to make informed decisions and contribute meaningfully to the societal conversations surrounding gene editing. Addressing misinformation and increasing public awareness about genetics and CRISPR technology is not merely a matter of scientific communication; it is a matter of building trust, enhancing transparency, and empowering individuals to actively participate in shaping the future of gene editing. By upholding ethical standards, nurturing a scientifically literate society, and fostering inclusive dialogues, we can pave the way for responsible innovation in gene editing while ensuring that the welfare and interests of all stakeholders are considered.

As we venture into the new frontiers of genetics, symbolized by gene editing and CRISPR technology, it is imperative that we address misinformation and increase public awareness. By engaging in effective communication, fostering inclusive conversations, prioritizing education, and promoting scientific literacy, we can overcome the challenges posed by misinformation and create a society that is well-informed, ethically aware, and actively involved in shaping the future of gene editing. Only through collective efforts can we unlock the full potential of these breakthroughs and ensure that they bring about positive change for humanity and the world we inhabit.

NURTURING PUBLIC TRUST AND ENGAGEMENT IN GENETIC RESEARCH AND ADVANCEMENTS

This is essential in order to ensure that the potential benefits of these breakthroughs can be fully realized. Genetic research and advancements have the power to revolutionize the medical field and improve the lives of individuals around the world. In order for this potential to be realized, it is crucial that the public has trust in the scientists and researchers working in this field and is actively engaged in the ethical and societal discussions surrounding genetic research. One of the key aspects of nurturing public trust is ensuring transparency in the research process. The public needs to have access to accurate and reliable information about genetic research and advancements. This information should be presented in a clear and understandable manner, free from scientific jargon, so that individuals can make informed decisions and form their own opinions about the technology. By providing the public with access to this information, researchers can demonstrate that they are acting in the best interests of society and are committed to ensuring the safe and responsible use of genetic technologies. In addition to transparency, fostering open dialogue and engagement with the public is also crucial. Genetic research and advancements have the potential to impact a wide range of individuals and communities, and it is important that their voices and opinions are heard and taken into account. By actively seeking input from the public, researchers can demonstrate that they value the perspectives of others and are committed to addressing the concerns and potential pitfalls associated with genetic research. This could include hosting public forums, engaging with community leaders and organizations, and creating opportunities for public input in the decision-making processes surrounding genetic research.

Building public trust requires strong ethical and regulatory frameworks. It is important for genetic research and advancements to be conducted within an ethical framework that respects the rights and dignity of individuals. This includes obtaining informed consent from individuals participating in research studies and ensuring that their privacy is protected. There need to be regulatory bodies in place to oversee genetic research and advancements, ensuring that they are conducted in a safe and responsible manner. Another important factor in nurturing public trust and engagement is addressing the potential risks and ethical dilemmas associated with genetic research and advancements. Genetic technologies such as CRISPR have the potential to be used for both beneficial and potentially harmful purposes. By acknowledging these risks and engaging in discussions about potential limitations and ethical considerations, researchers can demonstrate that they are committed to ensuring responsible use of these technologies. This could include exploring issues such as genetic enhancement, gene editing in human embryos, and the potential impact of genetic technologies on societal inequalities. In order to nurture public trust and engagement, there needs to be a commitment to ongoing education and awareness. Genetic research and advancements are complex and constantly evolving, and it is important that the public is kept informed about the latest developments in the field. This could include public education campaigns, workshops, and presentations designed to provide individuals with accurate and up-to-date information about genetic research and advancements. By equipping the public with knowledge and understanding, researchers can empower individuals to make informed decisions and actively engage in the discussions and debates surrounding this technology.

Nurturing public trust and engagement in genetic research and advancements is vital in order to fully realize the potential benefits of these breakthroughs. Through transparency, open dialogue, strong ethical frameworks, addressing risks and ethical dilemmas, and ongoing education and awareness, researchers can foster a sense of trust and engagement within the public. By actively involving the public in the decision-making processes surrounding genetic research, we can ensure that these advancements are conducted in a responsible and ethical manner and that the potential benefits are realized for all of humanity.

The field of genetics has undergone a revolution in recent years with the discovery and development of a powerful tool known as CRISPR. CRISPR is a gene editing technique that allows scientists to make precise changes to DNA. This breakthrough technology holds immense promise for the future of humanity in a number of ways. Firstly, CRISPR has the potential to cure a wide range of diseases that have long plagued humanity. By identifying and correcting specific genetic mutations that cause diseases such as cystic fibrosis, sickle cell anemia, and certain types of cancer, CRISPR offers the hope of eradicating these devastating conditions once and for all. In addition to disease treatment, CRISPR can also be utilized to enhance the genetic traits of organisms, bringing about a new era of genetically modified organisms (GMOs). Imagine crops that are more resistant to pests and diseases, animals with enhanced muscle growth for more efficient meat production, and even the ability to remove harmful genetic traits from human embryos. These possibilities have the potential to not only improve our quality of life but also address pressing global challenges such as food security and environmental sustainability. The future of humanity could indeed be shaped by the boundless possibilities of CRISPR.

One of the most exciting aspects of CRISPR is its ability to edit the human germline, meaning the changes made to an individual's DNA can be passed on to future generations. This raises both excitement and ethical concerns about the future of humanity. On one hand, the ability to eliminate genetic diseases from the gene pool and potentially enhance desirable traits offers great promise for improving the overall health and well-being of future generations. On the other hand, the potential for creating "designer babies" and the associated ethical and social implications poses significant challenges. Questions about what traits should be considered desirable and who gets to make those decisions are just some of the complex moral and ethical dilemmas that arise with the advent of CRISPR. As the technology continues to advance, it is crucial that we engage in thoughtful discussions and establish appropriate regulations to ensure the responsible use of gene editing to avoid unintended consequences and potential abuses. The implications of CRISPR extend beyond the scope of human genetics as well. By using CRISPR to alter the DNA of other organisms, we can potentially address pressing issues such as environmental degradation and the spread of disease. For example, with the ability to edit the genes of mosquitoes, which are carriers of diseases like malaria and dengue fever, we could potentially render them incapable of transmitting these deadly illnesses. Similarly, the modification of crops could lead to increased yields, improved nutritional content, and enhanced resistance to drought and pests. While the possibilities are vast, it is essential to consider the potential ecological and ethical consequences of such interventions. Balancing the benefits of gene editing with the preservation of biodiversity and ecosystem health will be key in shaping an environmentally sustainable future. Despite the immense potential of CRISPR, there are still significant challenges that must be overcome before its widespread application becomes a reality. One major obstacle is the off-target effects of CRISPR, meaning unintended changes to DNA that can lead to unforeseen consequences. Efforts are underway to minimize these off-target effects through improved delivery methods and more precise gene editing techniques. The ethical and societal implications of CRISPR demand careful consideration and regulation. An open and inclusive dialogue involving scientists, policymakers, and the wider public is necessary to navigate these complex issues and ensure responsible and equitable use of gene editing technologies. The discovery and development of CRISPR have opened up new frontiers in the field of genetics, offering the potential to cure diseases, modify organisms, and reshape the future of humanity. From the eradication of genetic diseases to the creation of genetically modified organisms, CRISPR holds immense promise for improving our quality of life and addressing pressing global challenges. The ethical and societal implications of gene editing cannot be overlooked, and it is crucial that we engage in thoughtful discussions and establish appropriate regulations to guide the responsible use of this powerful technology. As we journey to the edge of science, the possibilities and challenges presented by CRISPR will shape the future of humanity for generations to come.

XIV. FUTURE IMPLICATIONS AND POSSIBILITIES

The future implications of the groundbreaking advancements in genetics and CRISPR are boundless. With the ability to precisely modify an organism's genetic makeup, scientists hold the power to cure diseases that were once considered incurable. This has the potential to revolutionize healthcare and significantly improve the quality of life for millions of people around the globe. For instance, geneticists can now target and remove harmful mutations that cause debilitating genetic disorders, such as cystic fibrosis or Huntington's disease. By editing out these faulty genes and replacing them with healthy ones, scientists are opening up possibilities for treatments that were unimaginable just a few decades ago. The potential applications of CRISPR extend far beyond the realm of human health. Scientists have already begun using this technology to enhance agricultural crops, making them more resistant to pests and diseases. By editing the genes of these plants, scientists can create variations that are more robust, nutritious, and environmentally friendly. This not only holds the promise of solving global food shortages but also addresses the urgent need for sustainable agriculture practices in the face of a changing climate. CRISPR has the potential to reshape biodiversity. With this gene editing tool, scientists can control or eliminate invasive species that threaten ecosystems. By manipulating their genes, researchers can introduce sterility into these organisms or render them less fit for survival, ultimately preventing the damage and disruption they cause in their new habitats. In the same vein, CRISPR can be harnessed to protect endangered species by modifying their genome to enhance their adaptability or resistance to specific threats. This could offer a lifeline to many iconic species on the brink of extinction.

Another exciting future implication of CRISPR technology lies in the field of synthetic biology, where scientists engineer entirely new organisms with desired traits. This can range from bacteria that produce alternative fuels to biological sensors that detect and cleanse pollutants in the environment. By programming organisms with specific genetic instructions, scientists are poised to revolutionize industries such as energy, manufacturing, and waste management. The possibilities for custom-designed organisms to fulfill a wide range of human needs are limited only by our imagination and ethical considerations.

As with any powerful technology, there are also concerns and ethical questions that need to be addressed. The ability to edit the human germline DNA, for example, raises moral and ethical issues surrounding the modification of traits that could be inherited by future generations. The implications of "designer babies" and the potential for social inequality are topics that require careful consideration and regulation. There are concerns about the unintended consequences of genetic modifications, such as unforeseen ecological disruptions or the creation of unforeseen health risks. These concerns call for a well-informed and cautious approach toward the use of CRISPR and genetics in general.

The future implications and possibilities of genetics and CRISPR technology are both exciting and uncertain. From curing diseases to reshaping ecosystems and enhancing human capabilities, the potential for positive change is immense. This power comes with responsibility, and careful consideration must be given to the ethical, social, and ecological implications of these advancements. As we venture further into this new frontier of science, it is crucial that we engage in open dialogue and establish robust regulations to guide us toward a future where these technologies benefit humanity while respecting the values and diversity of our world.

ACCELERATING SCIENTIFIC PROGRESS IN GENE EDITING

The acceleration of scientific progress in gene editing has brought about a revolutionary era in genetics and holds immense potential for humanity. The advent of CRISPR system has streamlined the process of gene editing, allowing scientists to edit DNA with unprecedented precision and efficiency. This newfound capability has opened up a wide range of possibilities, from curing devastating diseases to altering organisms to benefit humanity in ways never thought possible. With each new breakthrough, the future of gene editing becomes increasingly promising, offering hope for a world free of genetic ailments and unimaginable advancements in various fields. One of the most significant advancements in gene editing is the potential to cure previously incurable genetic diseases. CRISPR has provided scientists with an effective tool to correct disease-causing genetic mutations in the DNA of affected individuals. By precisely targeting and modifying faulty genes, researchers can potentially eliminate the root cause of these debilitating ailments. For instance, CRISPR has shown promising results in treating genetic disorders like sickle cell disease and cystic fibrosis. These devastating conditions, which once seemed untreatable, now stand on the verge of a cure. Gene editing could pave the way for personalized medicine, where treatments are tailored to an individual's unique genetic makeup. This would revolutionize the field of healthcare, ensuring more effective and personalized treatments for patients around the world.

Beyond curing diseases, gene editing has the potential to reshape the very fabric of life itself. By altering the genetic code of organisms, scientists can create organisms with desired traits and characteristics. This opens up countless possibilities for improving agricultural crops, designing bacteria that produce valuable compounds, and even creating new species that are resilient to environmental changes. For instance, gene editing holds the potential to develop crops with increased resistance to pests and diseases, ensuring food security for millions of people. Similarly, scientists can engineer bacteria to produce renewable energy sources or life-saving drugs. The ability to rewrite the genetic code offers unprecedented power to shape the natural world according to human needs and desires. Gene editing has the potential to address pressing environmental concerns and mitigate the effects of climate change. By modifying the genes of organisms, scientists can potentially enhance their ability to adapt to changing environmental conditions. For example, researchers are exploring the possibility of genetically modifying coral reefs to make them more resistant to rising ocean temperatures and acidity levels. This could help preserve these vital ecosystems, which are under threat from climate change. Gene editing could be used to develop more sustainable biofuels by enhancing the efficiency of energy production in organisms. By harnessing the power of genetic engineering, humanity can mitigate the damage caused by climate change and safeguard the planet for future generations. While the acceleration of scientific progress in gene editing brings numerous opportunities, it also raises ethical concerns and challenges. The ability to modify the genetic code raises questions about the boundaries of human intervention in nature and the potential for unintended consequences. Gene editing technology carries the risk of off-target effects, where un-intended modifications occur in the genome, leading to unfore-seen complications. The permanent alteration of the germline, the genetic material passed on to future generations, raises eth-ical concerns about the potential for designer babies and the cre-ation of a genetically divided society. The accelerating scientific progress in gene editing, particularly with the advent of CRISPR technology, holds tremendous promise for humanity. The ability to cure genetic diseases, reshape organisms, and address press-ing environmental concerns has the potential to revolutionize various fields and reshape the future of humanity. Alongside these great advancements, ethical questions and challenges must be carefully considered to ensure the responsible and ethi-cal use of gene editing technology. With prudent regulation and a thoughtful approach, the new frontiers of genetics offer excit-ing possibilities that can usher in a brighter future for humanity.

UNLEASHING THE FULL POTENTIAL OF CRISPR TECHNOLOGY

The rapid development and refinement of CRISPR technology has opened up a world of possibilities in genetics and beyond. As researchers continue to explore the depths of this groundbreaking technique, it becomes increasingly apparent that the potential for disease eradication, organism modification, and the reshaping of humanity itself is within our reach. With the ability to precisely edit genes, CRISPR has the power to cure previously untreatable genetic diseases, offering hope to those afflicted and their families. This technology can be employed in creating disease-resistant organisms and crops, potentially revolutionizing the field of agriculture and ensuring food security for future generations. One of the most promising applications of CRISPR lies in its ability to cure genetic diseases that were once considered untreatable. Through the targeted editing of specific genes, scientists can potentially correct the underlying genetic mutations responsible for diseases such as Huntington's, cystic fibrosis, and sickle cell anemia. This precise gene editing holds the promise of not only alleviating the symptoms of these diseases but potentially eliminating the genetic disorder altogether, providing a permanent solution. This breakthrough has the potential to transform the lives of millions of individuals and families affected by these devastating conditions, offering new hope where there was previously only despair. CRISPR technology has the ability to modify organisms, bringing forth possibilities of creating disease-resistant plants and animals. By selectively editing genes responsible for susceptibility to diseases, scientists can create organisms that are naturally resistant, eliminating the need for chemical pesticides and antibiotics. This holds tremendous potential in the field of agriculture, where the creation of disease-resistant crops can increase yields and reduce the reliance on harmful chemicals. The modification of animal genes can lead to the production of disease-resistant livestock, ensuring healthier and more sustainable farming practices. The impact of such advancements in agriculture cannot be understated, as they have the ability to enhance food security and reduce the environmental strain caused by conventional farming methods.

The potential of CRISPR technology extends beyond curing diseases and modifying organisms. It has the capacity to fundamentally reshape the future of humanity. With the ability to edit the human germline, CRISPR allows for the alteration of genetic traits that can be inherited by future generations. While this frontier of genetic editing raises ethical questions and concerns, it also opens up exciting possibilities. It offers the potential to eliminate heritable genetic disorders and enhance desirable traits, such as intelligence or athleticism. This power also carries significant ethical implications, as decisions regarding which traits are desirable and who has access to gene editing raise complex societal and moral questions. In order to fully unleash the potential of CRISPR technology, careful regulatory frameworks must be put in place to ensure responsible and equitable use. The potential for misuse or unintended consequences demands caution and oversight. Striking a balance between scientific progress and ethical responsibility is crucial in navigating the uncharted waters of gene editing. Collaboration between scientists, policymakers, and the public is vital to ensure that the full potential of

CRISPR is realized in a manner that respects and upholds the principles of ethics and social justice.

The rapid advancement of CRISPR technology has brought us to the brink of a new era in genetics. From curing previously untreatable diseases to modifying organisms and reshaping the future of humanity, the possibilities are vast and exciting. With the ability to precisely edit genes, CRISPR offers hope to those afflicted by genetic disorders, while also holding the promise of revolutionizing agriculture and ensuring food security. Careful consideration and regulation are necessary to navigate the ethical implications of gene editing. As we venture deeper into the exciting world of genetics and CRISPR, it is essential that we proceed with both scientific rigor and ethical responsibility. Only then can we truly unleash the full potential of CRISPR and shape a future that benefits all of humanity.

WHAT LIES BEYOND THE HORIZON OF GENETICS?

Looking into the future of genetics, one cannot help but be captivated by the infinite possibilities that lie ahead. The field of genetics has seen tremendous advancements in recent years, particularly with the development of gene editing techniques such as CRISPR. With CRISPR, scientists have unlocked the ability to modify genetic material with unprecedented precision and ease, opening up a world of possibilities for the future of humanity.

In the realm of medicine, the potential of genetics is truly awe-inspiring. Gene editing has the potential to cure diseases that were once considered incurable. By altering the genetic code, scientists can correct the mutations responsible for genetic disorders, offering hope to millions of individuals and families affected by these conditions. This revolutionary approach to medicine has already shown promise in the treatment of diseases such as sickle cell anemia and cystic fibrosis, and the possibilities for future breakthroughs are limitless. Imagine a world where genetic diseases are eradicated, where individuals are no longer burdened by the limitations imposed by their genetic makeup. The future of medicine holds the promise of a healthier, disease-free world thanks to the power of genetics.

But the implications of genetic editing go beyond just medicine. The ability to modify genetic material opens up a whole new world of possibilities in agriculture and environmental conservation. By manipulating the genes of plants and animals, we could create new crop varieties that are more resistant to pests, drought, or other adverse conditions. This has the potential to increase food production and alleviate hunger in areas where resources are scarce. Gene editing could be used to preserve endangered species by enhancing their ability to adapt to changing environments or combating the genetic disorders that threaten their survival. The power to shape the genetic makeup of organisms offers us the opportunity to create a more sustainable and resilient world. Yet, with such immense power comes great responsibility. The ethical concerns surrounding gene editing are significant and cannot be overlooked. While the potential benefits are undeniable, we must tread carefully to avoid crossing ethical boundaries. The ability to genetically modify embryos, for example, raises questions about the moral implications of creating "designer babies" or selecting preferred traits. The line between improving the human condition and playing the role of a genetic architect is a fine one, and it is crucial that we engage in ongoing dialogue as a society to establish ethical guidelines.

Beyond the immediate future, the horizon of genetics holds even more breathtaking possibilities. As our understanding of the human genome deepens, we may one day have the ability to not only edit genes but also create entirely new ones. Imagine the potential for creating synthetic organisms with novel capabilities or enhancing our own genetic makeup to enhance human abilities. While these possibilities may still seem like science fiction, it is important to remember that what was once unimaginable quickly becomes reality in the rapidly advancing field of genetics. The boundaries of what we consider possible are constantly being pushed further, and the future holds the promise of truly mind-boggling advancements.

As we journey to the edge of science, it is vital that we approach the future of genetics with both excitement and caution. The potential of genetic editing has the power to transform our world in unimaginable ways, offering hope for disease eradication, food security, and environmental conservation. The ethical considerations that accompany these advancements must be carefully navigated to ensure that we do not lose sight of our shared humanity. The future of genetics is an adventure into uncharted territory, where the boundaries of what is possible are constantly being redefined. Let us take this journey with open minds and hearts, remembering that the decisions we make today will shape the future of humanity. The field of genetics has experienced a revolution in recent years with the advent of gene editing techniques such as CRISPR. This breakthrough technology has unlocked new frontiers previously unimaginable, offering potential cures for diseases, the ability to alter organisms, and the power to reshape the future of humanity. CRISPR is a gene-editing tool that allows scientists to modify specific genes within an organism's DNA. With CRISPR, geneticists can insert, delete, or replace specific pieces of DNA, providing a level of precision and efficiency never seen before. This newfound power has vast implications for treating genetic diseases, as scientists can now target and repair the underlying genetic mutations responsible for these disorders. By understanding the intricate code of life, we are entering a new era where we have the tools to rewrite it. One of the most promising applications of CRISPR is in the field of medicine. Up until now, treating genetic diseases has been a daunting task, as it required targeting and modifying specific genes responsible for the disorder. This process was time-consuming, expensive, and often ineffective. CRISPR has changed the game by offering a faster, cheaper, and more precise alternative. By using CRISPR, scientists can effectively edit an organism's DNA, correcting faulty genes and potentially curing genetic diseases. This advancement holds particular promise for conditions such as sickle cell anemia, Huntington's disease, and certain types of cancer. By directly addressing the genetic cause of these ailments, CRISPR offers a ray of hope for individuals who have long suffered from these debilitating conditions. The potential to alleviate human suffering and improve lives is truly revolutionary. CRISPR is not only useful for treating genetic diseases but also holds the potential to modify organisms, fueling advancements in agriculture and environmental conservation. With CRISPR, scientists can now edit the DNA of plants and animals, altering their genetic makeup to make them more resistant to diseases, pests, and adverse environmental conditions. This offers a promising solution to global food security challenges and the need for sustainable agriculture. By developing crops that have increased nutritional value, resistance to pests, or tolerance to harsh climates, we can address the growing demand for food without putting additional strain on our planet's resources. CRISPR allows us to explore the potential of de-extinction, bringing back extinct species by modifying the genetic code of related organisms. This exciting prospect could help restore biodiversity and undo some of the damage caused by human activity, offering a unique opportunity to reshape our natural world.

While the possibilities offered by CRISPR are vast, they also come with ethical dilemmas and concerns. The ability to edit genes raises questions about the potential for misuse or unintended consequences. With CRISPR, it is now possible to modify the genes of embryos, raising concerns about the ethics of gene editing for purposes such as selecting desirable traits or creating "designer babies". These questions touch upon fundamental issues of human identity, diversity, and the boundaries of scientific intervention. Ensuring responsible and ethical use of CRISPR technology is crucial to prevent misuse and the emergence of new inequalities in society. There are concerns about the potential for off-target effects, where unintended changes occur in an organism's DNA as a result of CRISPR. The long-term consequences of such alterations are still largely unknown and highlight the need for careful consideration and rigorous testing in the application of CRISPR technology. The field of genetics has been revolutionized by the advent of CRISPR. This breakthrough technology offers new possibilities for curing diseases, altering organisms, and shaping the future of humanity. By harnessing the power of this gene-editing tool, scientists can target and repair genetic mutations responsible for genetic diseases, potentially offering cures for previously incurable conditions. In addition, CRISPR holds promise for transforming agriculture, environmental conservation, and even bringing back extinct species. Ethical considerations and concerns about unintended consequences highlight the need for responsible use and further research. Despite these challenges, the exciting world of genetics and CRISPR offers a journey to the edge of science, presenting us with new opportunities and possibilities to improve human health, enhance food security, and reshape the natural world. Only time will tell how these new frontiers will shape our future, but one thing is certain genetic engineering and CRISPR are here to stay.

XV. CONCLUSION

The incredible advancements in genetics and the development of CRISPR technology have undoubtedly opened up new frontiers in the field of science and medicine. The ability to edit genes and manipulate DNA holds immense potential for curing diseases, improving crop yields, and even reshaping the future of humanity itself. As with any powerful tool, caution must be exercised to ensure that the ethical implications of gene editing are carefully considered and that potential negative consequences are actively addressed. Throughout this essay, we have examined the mechanics of CRISPR and the immense possibilities it offers for genetic editing. From the ability to eliminate genetic diseases before birth to the potential for enhancing desired traits in individuals, CRISPR has the potential to revolutionize the healthcare industry and improve the lives of millions. In addition, the application of CRISPR in agricultural practices has the potential to address the looming issue of food scarcity by improving crop yields and creating more robust and disease-resistant plants.

While the possibilities brought about by gene editing technology are certainly exciting, it is crucial to approach these advancements with a critical eye. The ethical implications of genetic manipulation cannot be ignored. On one hand, the ability to eradicate genetic diseases gives hope to countless individuals and families who have been burdened by these conditions for generations. On the other hand, the question of where to draw the line when it comes to genetic enhancement remains a pressing issue. The potential for creating a society where certain traits are prioritized over others raises concerns about discrimination and inequality. There are concerns about how gene editing may impact the natural diversity of species and ecosystems. By engaging in genetic modification, we risk disrupting delicate biological systems that have evolved over millions of years. While it may be tempting to believe that we possess the knowledge and understanding necessary to play the role of genetic architects, we must proceed with caution. Messing with nature, even with the best intentions, can have unintended consequences that are difficult to predict and undo. In order to navigate these complex ethical issues, it is essential that a well-informed and inclusive conversation takes place. Scientists, policymakers, ethicists, and the general public must come together to establish guidelines and regulations that ensure responsible use of these powerful tools. This involves not only considering the immediate consequences of gene editing, but also the long-term effects on society and the environment. There is a pressing need to address the potential for misuse or abuse of gene editing technology. As with any advancement in science, there is always the risk that it will be used for nefarious purposes, such as creating designer babies or developing bioweapons. Stringent regulations and oversight must be put in place to prevent these possibilities and to ensure that the benefits of gene editing are accessible to all, regardless of socioeconomic status or geographical location.

The new frontiers of genetics, particularly the development of CRISPR and gene editing technology, present us with a future full of unprecedented possibilities. We have the potential to cure previously incurable diseases, improve agricultural practices, and shape the future of humanity itself. With these possibilities come significant ethical implications and potential pitfalls that must be carefully considered. We have a responsibility to approach these advancements with caution, engaging in open and inclusive conversations and establishing regulations that prioritize the well-being of individuals, societies, and the natural world. Only through responsible utilization of these powerful tools can we ensure a future that brings positive change while minimizing the risks associated with gene editing technology. The journey to the edge of science continues, and it is our duty to tread carefully and responsibly as we explore these newfound frontiers.

THE POTENTIAL OF GENE EDITING AND CRISPR

Gene editing and CRISPR have the potential to revolutionize the field of genetics and reshape the future of humanity. These breakthroughs offer exciting possibilities in the field of medicine, as they hold the promise of curing diseases that were once considered incurable. With the ability to alter specific genes within an organism's DNA, scientists have the potential to treat genetic disorders by correcting the faulty genes responsible for the condition. This could potentially eliminate the need for invasive surgeries or lifelong medications, offering a more permanent and effective solution. Gene editing opens up the possibility of preventing genetic diseases altogether by editing the germline cells, which are responsible for passing on genetic traits to future generations. In this way, diseases that have plagued families for generations could be eradicated, leading to healthier future generations. Beyond medical applications, gene editing and CRISPR also hold tremendous potential in agriculture and livestock breeding. With the ability to modify the genetic traits of plants and animals, scientists could enhance crop yields, making them more resistant to pests, diseases, and environmental conditions. This could revolutionize agriculture, leading to increased food production and improved food security, especially in regions that are susceptible to droughts or other calamities. In the same vein, gene editing could also improve livestock and aquaculture, enhancing their resistance to diseases and improving the quality of meat and other animal products. These developments could not only increase the availability of affordable and nutritious food but also reduce the environmental impact of farming practices. Gene editing and CRISPR have the potential to unlock the mysteries of the natural world and further our understanding of various organisms. By manipulating genes, scientists can study the functions and interactions of specific genes within an organism, shedding light on their roles in development, behavior, and disease susceptibility. This knowledge could help us unravel complex genetic disorders in humans and pave the way for more effective treatments. Gene editing can be utilized to create animal models that mimic human diseases, allowing scientists to study these conditions in a controlled environment and develop potential therapeutics. These advancements could profoundly impact our understanding of biology and contribute to the development of new drugs and treatments. While the potential of gene editing and CRISPR is vast, it is crucial to approach these technologies with caution and ethical considerations. The ability to modify genes raises important ethical questions regarding the extent to which we should intervene in the natural order of life. There are concerns about unintended consequences and the potential for misuse of these technologies. Genetic modifications made to an organism's DNA could have unintended effects on other genes or even lead to unforeseen environmental consequences. As such, it is imperative for scientists and policymakers to proceed with rigorous testing and regulation to ensure the responsible and safe use of gene editing. Gene editing and CRISPR represent a new frontier in the field of genetics that holds immense potential for medicine, agriculture, and scientific research. These breakthroughs offer the promise of curing genetic diseases, improving food production, and enhancing our understanding of the natural world. It is essential to proceed with caution and carefully consider the ethical implications and potential risks associated with these technologies. Both scientific rigor and ethical considerations should guide our exploration of gene editing and CRISPR in order to ensure the responsible and beneficial use of these revolutionary tools. Only through responsible and thoughtful application can we fully harness the potential of gene editing and CRISPR without compromising the integrity of life itself.

IMPACT ON CURING DISEASES, ALTERING ORGANISMS, AND RESHAPING HUMANITY

The impact of gene editing, specifically through CRISPR technology, has the potential to revolutionize the field of medicine by offering new avenues for curing diseases, altering organisms, and reshaping humanity. Firstly, the ability to edit the genetic code of human beings presents a promising opportunity for combating and even eradicating debilitating diseases. With the precision and accuracy that CRISPR allows, scientists can target and repair specific genetic mutations that contribute to various illnesses. For instance, diseases like sickle cell anemia and cystic fibrosis, which are caused by single gene mutations, could potentially be cured through gene editing. By modifying the defective gene responsible for these conditions, it is possible to restore normal functioning and alleviate the suffering of those affected. CRISPR offers a means to combat complex diseases such as cancer, where multiple genes are involved. By manipulating key genes or introducing new ones, scientists can reprogram cancer cells to self-destruct or enhance the body's natural defenses against the disease. This breakthrough technology provides a glimmer of hope for countless individuals plagued by these devastating illnesses. Beyond the realm of human diseases, gene editing has the potential to alter organisms in profound ways. Agricultural practices, for example, can be transformed through the application of CRISPR to produce crops that are more resistant to pests and diseases, or that possess greater nutritional value. By introducing desired traits into the genetic makeup of plants, scientists can enhance crop productivity and increase food security, addressing crucial issues in a world grappling with overpopulation and food scarcity. Similarly, gene editing can be used in livestock breeding to produce animals with improved traits such as disease resistance or higher meat yield. This not only benefits the agricultural industry but also holds the promise of reduced environmental impact and more sustainable farming practices.

The impact of gene editing technology extends beyond the realm of medicine and agriculture. It has the potential to reshape humanity in ways previously unimaginable. The ability to alter the genetic makeup of humans raises profound ethical questions and challenges societal norms. CRISPR allows for the modification of germline cells, which means that the alterations made to an individual's genetic code can be passed on to future generations. This raises concerns about the creation of "designer babies" and the potential for widening socioeconomic disparities. The notion of editing specific traits raises moral dilemmas surrounding ideas of perfection and the potential erasure of diversity. A futuristic world where certain traits are sought after and others are deemed undesirable may have far-reaching implications for the concept of human identity and the value we place on individual differences. Despite these ethical concerns, the potential benefits of gene editing technology are undeniable. It offers a window into a future where genetic diseases can be cured, crops can be enhanced, and humanity can be shaped in ways that were once reserved for science fiction. It is crucial to proceed with caution and adhere to strict ethical guidelines to ensure the responsible use of this technology. Regulatory frameworks must be established to govern the use of gene editing and extensive research is needed to fully understand the long-term effects and potential risks associated with these interventions.

Gene editing technology, particularly through the use of CRISPR, has the potential to have a transformative impact on the world. The ability to cure diseases, alter organisms, and reshape humanity brings with it both great promise and ethical considerations. By harnessing the power of gene editing, we have the opportunity to alleviate human suffering through the eradication of genetic diseases, enhance agricultural practices to combat food scarcity, and adapt humanity to meet the challenges of an ever-changing world. The new frontiers of genetics offer exciting possibilities, but they must be navigated responsibly to ensure the best outcomes for all of humanity.

FINAL THOUGHTS ON THE NEED FOR RESPONSIBLE AND ETHICAL APPLICATION OF GENE EDITING TECHNOLOGIES

The remarkable advancements in gene editing technologies, particularly CRISPR, present an incredible opportunity for scientific progress and a potential revolution in the field of genetics. The ability to modify genes and potentially cure diseases that were once deemed incurable is undoubtedly a cause for excitement and optimism. It is paramount that these technologies are used responsibly and ethically, ensuring that we consider the potential consequences and implications of our actions.

One of the key concerns surrounding the use of gene editing technologies is the potential for unintended consequences and unintended biases. Although the precision and efficiency of CRISPR are impressive, there is still the potential for off-target effects, which could lead to unintended alterations in the genome. This possibility raises ethical concerns, as altering the genome can have irreversible effects on an individual and potentially future generations. It is crucial that rigorous safety protocols and testing are in place to minimize the risk of unintended consequences.

The responsible use of gene editing technologies necessitates an open dialogue about the implications and ethical considerations associated with these advancements. It is essential to engage diverse stakeholders, including scientists, policymakers, and the public, in discussions regarding the acceptable limits of gene editing, potential risks, and the societal impact of these technologies. The voices of marginalized communities and indigenous peoples, who may be disproportionately affected by gene editing technologies, must be included to ensure equitable decision-making. Ethical guidelines and regulations should be established to guide the responsible implementation of these technologies, promoting transparency, accountability, and inclusivity.

The potential use of gene editing technologies for non-medical purposes raises ethical concerns. The ability to modify the genetic traits of organisms, such as crops or livestock, opens up a range of possibilities, from enhancing desirable traits to potentially creating designer babies. While the enhancement of certain traits may seem desirable in some contexts, it also raises questions about the commodification of life and the potential for exacerbating existing inequalities. It is crucial to carefully consider the ethical implications of such applications and establish safeguards to prevent the misuse or abuse of gene editing technologies. Another important consideration is the potential impact on the natural world. The release of genetically modified organisms into the environment could have unintended ecological consequences. The responsible application of gene editing technologies must include thorough risk assessments and precautions to prevent the unintended spread of genetically modified organisms and potential ecological disruptions.

It is essential to address the ethical concerns related to the accessibility and equity of gene editing technologies. The high costs associated with gene editing techniques could potentially exacerbate existing inequalities, limiting access to those who can afford it while further marginalizing disadvantaged communities. Any responsible application of gene editing technologies must also include measures to ensure equitable access, affordability, and the prevention of potential disparities in healthcare.

A responsible and ethical approach to gene editing technologies necessitates an ongoing evaluation and monitoring of their long-term effects. The potential impact of genetic modifications on future generations and the wider population must be continually studied and assessed. Long-term safety studies and post-imple-mentation monitoring are essential to identify any unforeseen consequences or risks and adjust our usage accordingly.

The development of gene editing technologies, especially the revolutionary CRISPR system, holds remarkable promises for the future of humanity. It is crucial that we approach these technol-ogies with a sense of responsibility and ethics. Rigorous safety protocols, inclusive decision-making processes, and ethical guidelines are necessary to ensure the responsible application of gene editing technologies. An ongoing evaluation of their effects, both short-term and long-term, is also necessary to mitigate any unintended consequences. By striving for responsible and ethical practices, we can ensure that the potential benefits of gene ed-iting are maximized while minimizing the risks and potential harms associated with these powerful tools. Only through thoughtful and deliberate action can we fully embrace the po-tential of gene editing and pave the way for a future in which these technologies are harnessed for the betterment of humanity.

BIBLIOGRAPHY

Dr. Chetana V. Donglikar , Dr. Rajabhau C. Korde. 'Challenges and Opportunities of National Education Policy 2020 Before Higher Education.' Ashok Yakkaldevi, 4/20/2023

Posetti, Julie. 'Journalism, fake news & disinformation.' handbook for journalism education and training, Ireton, Cherilyn, UNESCO Publishing, 9/17/2018

Heather K Allen. 'Virus Ecology and Disturbances: Impact of Environmental Disruption on the Viruses of Microorganisms.' Stephen Tobias Abedon, Frontiers Media SA, 3/26/2015

Nicholas Steiner. 'Unforeseen Consequences.' Xlibris Corporation, 12/1/2006

John Strain. 'Ethics for Living and Working.' Simon Robinson, Troubador Publishing Ltd, 1/1/2008

National Academy of Sciences. 'Heritable Human Genome Editing.' The Royal Society, National Academies Press, 1/16/2021

Tiit Tammaru. 'Urban Socio-Economic Segregation and Income Inequality.' A Global Perspective, Maarten van Ham, Springer Nature, 3/29/2021

Food and Agriculture Organization of the United Nations . 'The impact of disasters and crises on agriculture and food security: 2021.' Food & Agriculture Org., 3/17/2021

David B. Collinge. 'Plant Pathogen Resistance Biotechnology.' John Wiley & Sons, 4/8/2016

Sajid Fiaz. 'Principles and Practices of OMICS and Genome Editing for Crop Improvement.' Channa S. Prakash, Springer Nature, 7/18/2022

Mark Overton. 'Agricultural Revolution in England.' The Transformation of the Agrarian Economy 1500-1850, Cambridge University Press, 4/18/1996

David L. Hawksworth. 'GMOs.' Implications for Biodiversity Conservation and Ecological Processes, Anurag Chaurasia, Springer Nature, 11/30/2020

Institute of Medicine. 'Safety of Genetically Engineered Foods.' Approaches to Assessing Unintended Health Effects, National Research Council, National Academies Press, 7/8/2004

Jonathan Wolff. 'Risks and Regulation of New Technologies.' Tsuyoshi Matsuda, Springer Nature, 12/1/2020

C.A. Brebbia. 'Environmental Impact II.' G. Passerini, WIT Press, 5/14/2014

Health and Medicine Division. 'Communities in Action.' Pathways to Health Equity, National Academies of Sciences, Engineering, and Medicine, National Academies Press, 4/27/2017

Institute of Ideas. 'Designer Babies.' Where Should We Draw the Line?, Ellie Lee, Hodder & Stoughton, 1/1/2002

Muzaffar Munshi. 'Biohacking: How Technology is Changing Our Bodies.' Muzaffar Munshi, 5/13/2023

Erik Parens. 'Enhancing Human Traits.' Ethical and Social Implications, Georgetown University Press, 1/1/1998

Patrick Arbuthnot. 'Gene Therapy for Viral Infections.' Academic Press, 6/1/2015

Life Extension. 'Disease Prevention and Treatment.' Life Extension, 1/1/2013

Edward Tenner. 'Why Things Bite Back.' Technology and the Revenge of Unintended Consequences, Knopf, 1/1/1996

C. David Coats. 'Old MacDonald's Factory Farm.' The Myth of the Traditional Farm and the Shocking Truth about Animal Suffering in Today's Agribusiness, Crossroad Publishing Company, 1/1/1991

Division on Earth and Life Studies. 'Genetically Engineered Crops.' Experiences and Prospects, National Academies of Sciences, Engineering, and Medicine, National Academies Press, 1/28/2017

Theodore Friedmann. 'Gene Therapy.' Fact and Fiction in Biology's New Approaches to Disease, CSHL Press, 1/1/1994

National Academy of Medicine. 'Human Genome Editing.' Science, Ethics, and Governance, National Academies of Sciences, Engineering, and Medicine, National Academies Press, 8/13/2017

Shiu-Jau Chen. 'Gene Editing.' Technologies and Applications, Yuan-Chuan Chen, BoD – Books on Demand, 5/29/2019

Bruce Alberts. 'Molecular Biology of the Cell.' Garland, 1/1/2004

Alfred Henry Sturtevant. 'A History of Genetics.' CSHL Press, 1/1/2001

New York-Mid-Atlantic Consortium for Genetic and Newborn Screening Services. 'Understanding Genetics.' A New York, Mid-Atlantic Guide for Patients and Health Professionals, Genetic Alliance, Lulu.com, 1/1/2009

John van der Oost. 'CRISPR-Cas Systems.' RNA-mediated Adaptive Immunity in Bacteria and Archaea, Rodolphe Barrangou, Springer Science & Business Media, 12/13/2012

Christoph Lange. 'The Fundamentals of Modern Statistical Genetics.' Nan M. Laird, Springer Science & Business Media, 12/13/2010

www.ingramcontent.com/pod-product-compliance
Lightning Source LLC
Chambersburg PA
CBHW072356290526
45794CB00001B/90